Yasuto Itoh

Gunchu Formation

Yasuto Itoh

Gunchu Formation

An Indicator of Active Tectonics on an Oblique Convergent Margin

LAP LAMBERT Academic Publishing

Impressum / Imprint

Bibliografische Information der Deutschen Nationalbibliothek: Die Deutsche Nationalbibliothek verzeichnet diese Publikation in der Deutschen Nationalbibliografie; detaillierte bibliografische Daten sind im Internet über http://dnb.d-nb.de abrufbar.

Alle in diesem Buch genannten Marken und Produktnamen unterliegen warenzeichen-, marken- oder patentrechtlichem Schutz bzw. sind Warenzeichen oder eingetragene Warenzeichen der jeweiligen Inhaber. Die Wiedergabe von Marken, Produktnamen, Gebrauchsnamen, Handelsnamen, Warenbezeichnungen u.s.w. in diesem Werk berechtigt auch ohne besondere Kennzeichnung nicht zu der Annahme, dass solche Namen im Sinne der Warenzeichen- und Markenschutzgesetzgebung als frei zu betrachten wären und daher von jedermann benutzt werden dürften.

Bibliographic information published by the Deutsche Nationalbibliothek: The Deutsche Nationalbibliothek lists this publication in the Deutsche Nationalbibliografie; detailed bibliographic data are available in the Internet at http://dnb.d-nb.de.

Any brand names and product names mentioned in this book are subject to trademark, brand or patent protection and are trademarks or registered trademarks of their respective holders. The use of brand names, product names, common names, trade names, product descriptions etc. even without a particular marking in this work is in no way to be construed to mean that such names may be regarded as unrestricted in respect of trademark and brand protection legislation and could thus be used by anyone.

Coverbild / Cover image: www.ingimage.com

Verlag / Publisher:
LAP LAMBERT Academic Publishing
ist ein Imprint der / is a trademark of
OmniScriptum GmbH & Co. KG
Heinrich-Böcking-Str. 6-8, 66121 Saarbrücken, Deutschland / Germany
Email: info@lap-publishing.com

Herstellung: siehe letzte Seite /
Printed at: see last page
ISBN: 978-3-659-39898-8

Table of Contents

Preface

The purpose of this book is to elucidate the intriguing tectonics of an active plate margin based upon authentic field geology. Recent quantum leaps in the field of geophysical technology, such as reflection seismic surveys and seismic tomography, have provided us with precise three-dimensional views of the Earth's deep interior. It is especially useful to disclose the complicated development process of active margins. In order to comprehensively understand the geodynamics, however, describing surface processes of the globe by means of the field-based stratigraphy, sedimentology and structural geology still has equal significance because such geologic approaches suggest the time sequence of tectonic events and provide us with a foresight into coming changes.

As a case study, the author presents detailed geologic data for an event sedimentary unit on a large transcurrent active fault, the Median Tectonic Line, bisecting the southwestern Japan arc. Compositional clastic changes of the unit demonstrate a complex history of uplift and exhumation of the hinterlands, and radiometric age data using the latest methods give insight into sediment provenance, depicting formation and migration processes of a conspicuous asymmetric basin.

Thus the sedimentation process has been controlled by the morphological features of a fault-associated basin. In conjunction with a regional tectonic framework, unique structural development of the unit under study indicates possible Quaternary changes in the fault-slip sense, which is closely linked with the mode of oceanic plate convergence having direct impact on deformation modes and paleoenvironment of the island arc.

Therefore, a well-organized geologic investigation within a confined area eventually contributes to general geodynamic theory given that multidisciplinary information can be suitably integrated.

Yasuto Itoh
Graduate School of Science,
Osaka Prefecture University
Japan

1. Introduction

Convergent plate margins are accompanied by vigorous deformations and the emergence of island arcs under strong tectonic stress. They are sites where a variety of basins form, reflecting large geomorphic diversity. Obliqueness of the convergence can result in activity on an arc-bisecting fault (Fitch, 1972), compartmentalization of sedimentary basins (Takano et al., 2013), and eventual lateral transportation and massive rearrangement of the constituents along an active margin.

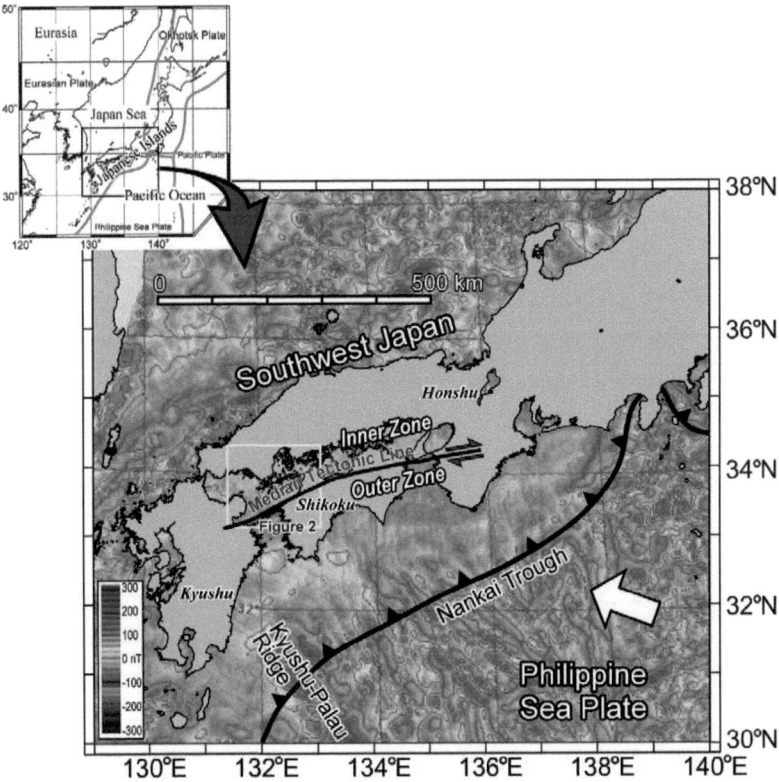

Figure 1. Index map showing tectonic regimes around the eastern Eurasian margin. Magnetic anomaly map in the ocean is after Geological Survey of Japan and CCOP (2002). An open arrow shows the present relative motion of the subducting Philippine Sea Plate.

The Japanese Archipelago (Figure 1) is a typical arc-trench system, and its geologic architecture has been studied since the 19th century from various viewpoints. Plate tectonics established in the latter half of the 20th century clarified the causal relationship between the theory of continental drift and orogenic movements. As for the circum-Pacific region, a chronicle of plate motion over the last 100 million years was presented by Engebretson et al. (1985). Northeast Japan is under the influence of longstanding subduction of the Pacific Plate, which has provoked subduction erosion in the forearc region and intra-forearc basin formation. On the other hand, southwest Japan is characterized by the growth of an accretionary prism from the middle Mesozoic to early Cenozoic.

Development of southwest Japan as an island arc has been controlled by a complicated interaction of converging plates (Philippine Sea, Pacific, Okhotsk or North American) since backarc rifting, which led to the formation of the current plate configuration. Formation of the Japan Sea, the backarc basin of southwest Japan, took place in a short period around the Early Miocene according to a series of pioneering paleomagnetic research studies by Otofuji and his colleagues (e.g., Otofuji and Matsuda, 1987; Otofuji et al., 1985). As shown in Figure 1, the geomagnetic anomaly pattern in the basin floor lacks a clear spreading center, which is attributed to divergent rifting in the southern Japan Sea (Itoh et al., 2006), as deduced from precise paleoreconstruction.

Since the mid-Cenozoic, the forearc region of southwest Japan has been under the influence of subduction of the Philippine Sea Plate. The Shikoku Basin, a component of the oceanic plate converging on the Nankai Trough, is accompanied by geomagnetic lineation (Figure 1; Kobayashi and Nakada, 1978; Seno and Maruyama, 1984), and has been assigned to 24~15 Ma (Okino et al., 1994). The Philippine Sea Plate has large topographic relief such as the remnant arc of the Kyushu-Palau Ridge (Figure 1), which may provoke confined strong deformation of the overriding plate margin.

Nakamura et al. (1987) described the submarine topography and shallow structure along the eastern forearc slope of southwest Japan. They advocated a counterclockwise shift of the convergent direction of the movement of the Philippine Sea plate from north-northwestward to its present west-northwestward motion (open arrow in Figure

1), calculated based on the work of Seno et al. (1993), reckoning that the shift occurred around 2~1 Ma. The highly oblique convergence of the oceanic plate reactivated dextral slips on the Median Tectonic Line (hereafter abbreviated as MTL), which divides the northern Mesozoic granitic terrane (Inner Zone) and the southern accretionary complex (Outer Zone) of southwest Japan. Prominent shear stress has caused wrench deformation of the Inner Zone (Itoh and Takemura, 1993) and eventual crustal breaking along the coast of the Japan Sea (Itoh et al., 2002).

Sedimentary basins serve as a sensitive indicator of tectonic processes that can aid in a comprehensive understanding of such complicated phenomena. As shown by Itoh (2013), the spatiotemporal distribution of basins suggests progressive deformation of an island arc, and their infilling sediments record uplift and exhumation of hinterlands and a mass balance along a convergent margin. Among his edited collections, Itoh et al. (2013) argued a characteristic basin formation process at the terminations of the MTL, mainly on the basis of gravity and geomagnetic data. In this book, the author concentrates on the northwestern portion of Shikoku Island (Figures 1, 2 and 3), where a specific Quaternary sedimentary unit, the Gunchu Formation, rests on the MTL and records active tectonics linked to changes in the convergence mode of the Philippine Sea Plate. The author aims to quantify the tectonics on an active plate margin based on authentic field geology and various kinds of physical analyses.

2. Review

2.1. Geophysics

Marine gravity surveys have been conducted in the Seto Inland Sea surrounded by the Honshu, Kyushu and Shikoku Islands of southwest Japan (Figure 1) in recent decades (Yusa et al., 1992; Koizumi et al., 1994; Ohno et al., 1994; Kono et al., 2001). They revealed an extensive negative anomaly zone around the Iyonada Sea, in the western part of the Inland Sea. Yusa et al. (1992) originally tried to construct a subsurface structural model that fitted the anomaly trend adjacent to the eastern coast of Kyushu. Ohno et al. (1994) expanded the modeling area to the entire Iyonada Sea (Figure 2). Although these studies assumed three units with different densities in the basement rock, its geologic context still remains ambiguous.

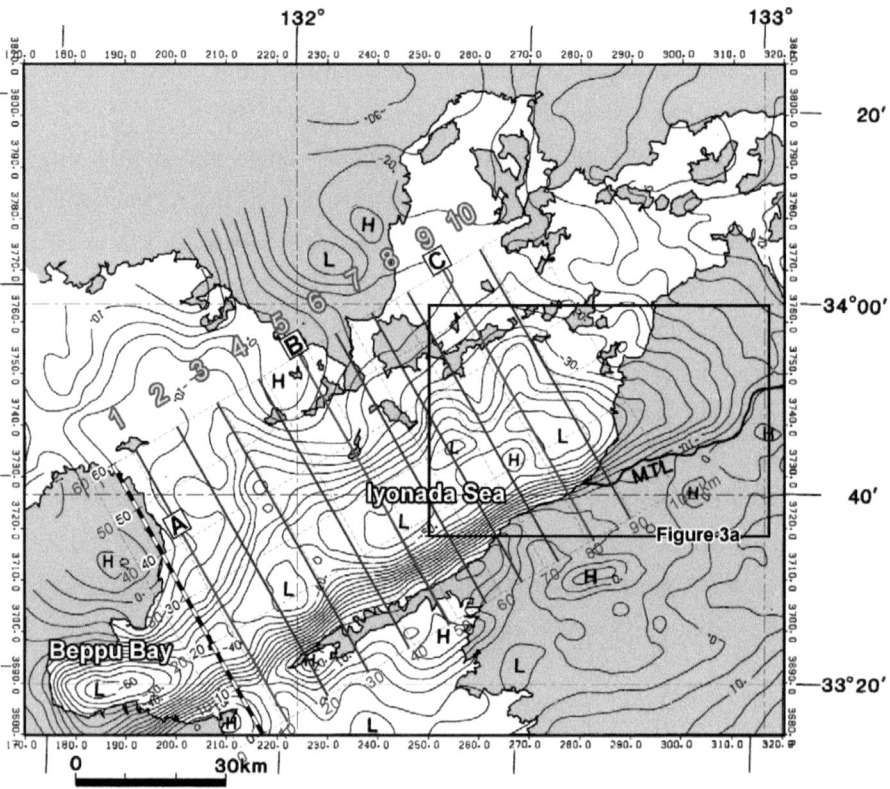

Figure 2. Bouguer gravity anomaly trend around the western Setouchi (Seto Inland Sea) Province (Ohno et al., 1994). The Bouguer density is 2670 kg/m^3. The contour interval is 5 mGal. Lines A to C and 1 to 10 are for 2-D gravity anomaly modeling in the Iyonada Sea after Ohno et al. (1994) and Itoh et al. (2013), respectively.

Itoh et al. (2013) adopted a simpler two-layered (sediment and basement) model. They applied the method of Talwani et al. (1959) to estimate two-dimensional subsurface structures for 10 modeling lines. Their model of the remarkable elongate

8

depression has a profile common with a half-graben, implying a tensile stress state. Itoh et al. (2013) suggested that an isopach map for the top basement structure reaches 4 km from mean sea level at the deepest part of the basin. The volume of the basement depression was calculated as 7.2×10^3 km^3 using Gauss-Legendre numerical integration (Davis and Rabinowitz, 2007) with depth data on a 10 km interval mesh. Assuming that the basin-filling clastics were supplied from an adjacent hinterland in the northwestern quarter of Shikoku, this total volume requires average erosion 1500 m thick. This is a sufficient thickness to have an impact on regional isostasy, and using age determination of the thick pile of sediments is an inevitable step in elucidating the tectonic processes and mass balance on the plate boundary.

2.2. Geology

The geologic system in the study area, which faces the eastern Iyonada Sea (Figure 3), is summarized in Figure 4 after Takahashi et al. (1990). It is known that a conspicuous sedimentary unit known as the Gunchu Formation is distributed along the northwestern coast of Shikoku (Nagai, 1957). It is a non-marine deposit containing abundant plant remains, and has a steep homoclinal structure affected by the Quaternary activity of the MTL running along the southern margin of the Iyonada Sea (Ogawa et al., 1992). The middle part of the Gunchu Formation contains a considerable amount of crystalline schist gravels derived from the Sanbagawa metamorphic belt in the Outer Zone (Mizuno, 1987).

Details of the stratigraphy and the geologic structure of the Gunchu Formation were reported by Takahashi and Kashima (1985). Mizuno (1987) reviewed previous geologic studies and presented columnar sections with horizons of volcanic ash layers. Itoh et al. (2013) presented the results of a preliminary sedimentological survey and chronological analysis of coastal outcrops of the Gunchu unit. These surveys were rather restricted to the well-exposed coastal section, and data coverage in the interior areas remained low at the beginning of the present study.

As for stratigraphic correlation, Mizuno (1992) showed a synthesis of the regional

9

correlation of the Pleistocene series around the Seto Inland Sea based upon tephrochronology, and posited basin-forming processes. It seems, however, that radiometric dates of key volcanic ashes should be reexamined using latest methods because large scatter and uncertainty in numerical ages hinder to determine precise sequence of basin formation events.

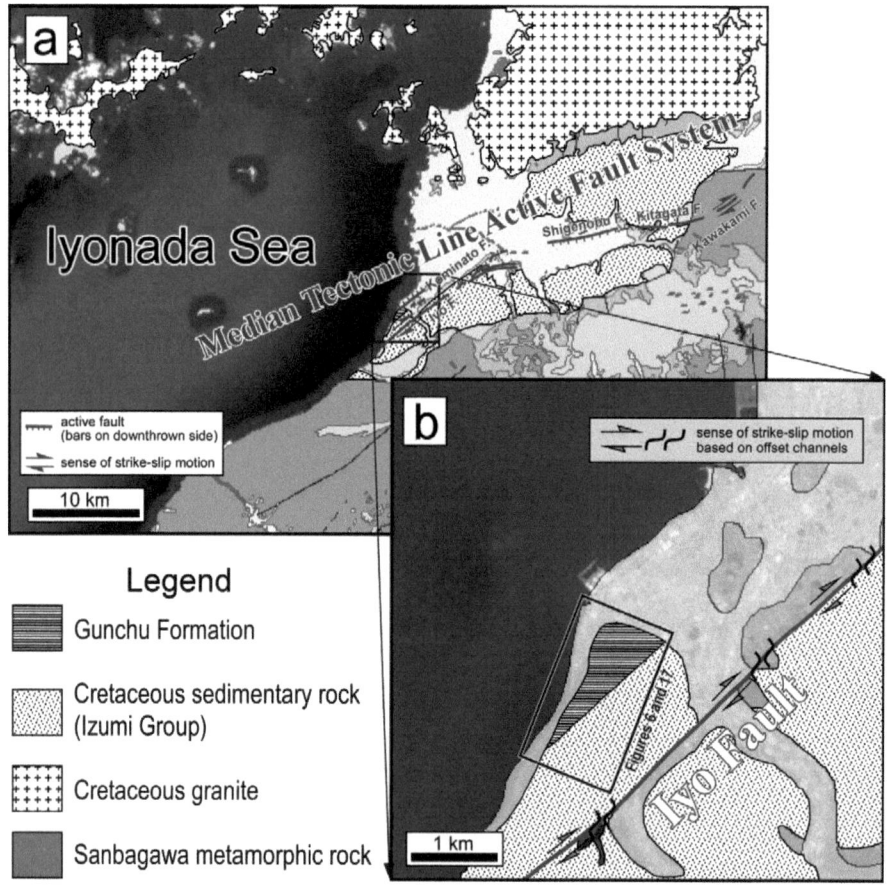

Figure 3. Simplified geology of the western Setouchi Province. (a) Distribution of the Median Tectonic Line active fault system is after Nakata and Imaizumi (2002). The base map is after Geological Survey of Japan (2012). (b) Sense of strike-slip motion annotated on the Iyo Fault is after Okada et al. (1998).

Holocene	*Alluvium*	
Pleistocene	*Gunchu Formation*	nonmarine clay, silt, sand, gravel and their alternations
Late Cretaceous	*Izumi Group*	marine alternation of sandstone and mudstone

Figure 4. Geologic system in the Iyo area.

2.3. Tectonics

As mentioned in the previous section, Quaternary oblique subduction of the Philippine Sea Plate revitalized a long-lived crustal break as the Median Tectonic Line active fault system (hereafter referred to as MTLAFS). The active Kominato, Hongu and Iyo Faults have been identified around the area of the Gunchu Formation (Nakata and Imaizumi, 2002). Among them, clear dextral motion is confirmed on the Iyo Fault based on channel offsets (Figure 3b; Okada et al., 1998), and the Iyo Fault is regarded as a significant component of the MTLAFS.

Transition of active segments in the MTLAFS should have provoked formation of pull-apart basins. Itoh et al. (1998, 2014) clarified the basin-forming process based on densely arrayed reflection seismic data from off the eastern shore of central Kyushu (Beppu Bay in Figure 2). Because of a lack of seismic information, the architecture of the adjoining Iyonada Sea still remains unsolved.

Ikeda et al. (2009) argued a tectonic model and fault segmentation of the MTLAFS within Shikoku on the basis of gravity and reflection seismic data. They pointed out contrasting stress-strain states along the regional fault, namely, transpression and transtension structural development in eastern and western Shikoku, respectively. Their interpretation was that the conspicuous deformation mode could be attributed to counterclockwise motion of the forearc sliver relative to the mother continent, based on GPS observations (Nishimura and Hashimoto, 2006). On the other hand, a detailed geologic review by Itoh et al. (2014) suggested that larger areas had been involved in the tectonic episode, implying a more regional motive force. They assumed that such a tectonic trend was related to lopsided subsidence, namely a jump of the Euler pole of the Philippine Sea/Eurasian Plates. This working hypothesis will be upgraded through discussion in this book based on geologic evidence.

3. Geologic survey

The Gunchu Formation was deposited under an entirely nonmarine environment, filling a Pleistocene alluvial basin. It consists of voluminous coarse clastics in channel-levee systems intercalated by fine deposits in a floodplain or back marsh. Fine sediments of the Gunchu Formation yield abundant plant remains such as *Picea maximowiczii, Pseudolarix kaempferi, Metasequoia glyptostroboides, Chamaecyparis pisifera, Juglans megacinerea, Pterocarya multistriata* and *Hamamelis parrotioidea* (Mizuno, 1987).

Mizuno (1987) divided the succession into the Lower, Middle and Upper Members following the lithologic description of Takahashi and Kashima (1985). Minor sequential changes in sedimentary facies reflect migration of meandering rivers on the alluvial plain, and the unit basically represents one coarsening-upward cycle generated by increases in sediment delivery, which is linked to tectonic uplift and exhumation of hinterlands.

There is controversy about the relationship between the Gunchu Formation and the underlying Cretaceous Izumi Group. Takahashi and Kashima (1985) argued that the units have a fault contact, surmised from observation of brecciated sandstone of the Izumi Group. On the other hand, Mizuno (1987) pointed out that the geologic boundary is not accompanied by a clear lineament, implying fault activity, and stated that the Gunchu Formation rests unconformably upon the undulate surface of the Izumi Group.

The author has conducted a total of two months of field surveys of the Gunchu Formation since 2012, and has confirmed the points listed below:

Mizuno (1987)	Present Study
Upper Member Gravel-dominant unit intercalating many coaly sand and silt layers. - - - - - - - - - - - - - - - - - - **Middle Member** Gravel (abundant schist clasts)-dominant unit intercalating many sand, silt and clay layers.	**Upper Member** Gravel-dominant unit. Metamorphic clasts and sand clasts are abundant in its lower part and upper part, respectively.
Lower Member Alternation of gravel and silt-clay intercalated by coaly sand layers. It yields nonmarine mollusca and plant remains.	**Lower Member** Alternation of gravel, sand and silt intercalating coaly clay layers. Plant remains are abundant. This unit is characterized by presence of granitic material in gravel beds.

Figure 5. Stratigraphic summary of the Gunchu Formation.

1) From an evolutionary process viewpoint of the sedimentary basin, the succession should be divided into the Lower and Upper Members (Figure 5).

2) The Gunchu Formation has an unconformable relationship with the underlying Cretaceous Izumi Group.

3) Four fault blocks are identified in the area where the Gunchu Formation is distributed based on offsets of stratigraphic unit boundaries (Figure 6).

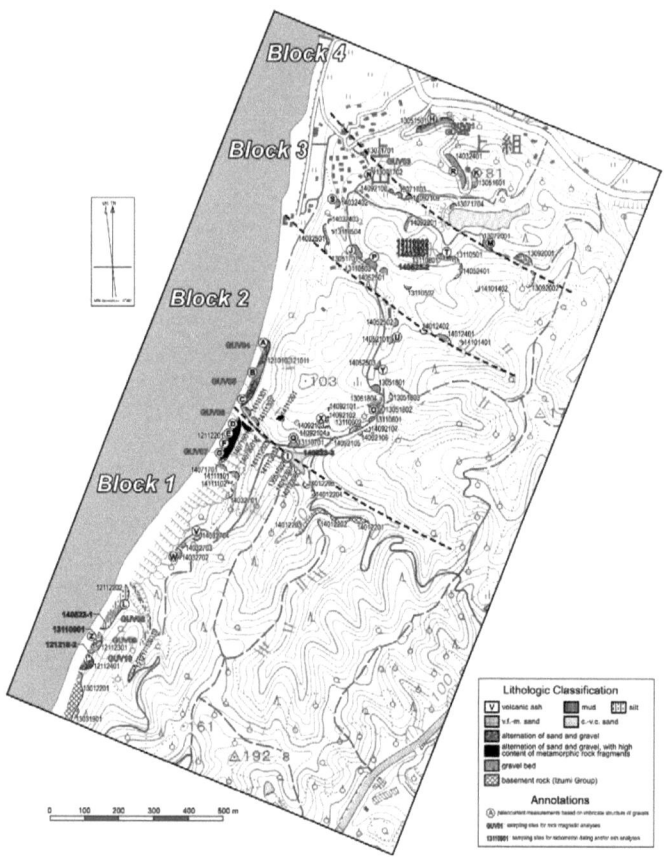

Figure 6. Lithologic map of the Gunchu Formation. The base map is a part of the "Kaminada" 1:25,000 topographic map published by the Geographical Survey Institute of Japan.

4) Reflecting active tectonics, a drastic change occurred in clastic provenance during the deposition of the alluvial unit, which is shown by gravel compositions and paleocurrent directions measured using the imbricate structure of the gravels.

Figures 7a to 7d present columnar sections for each fault block, together with the dominant clast types of the gravel beds and sampling horizons for various analyses. The locations of the columns are shown in Figure 6. Detailed results of the field survey are described in the following sections.

14

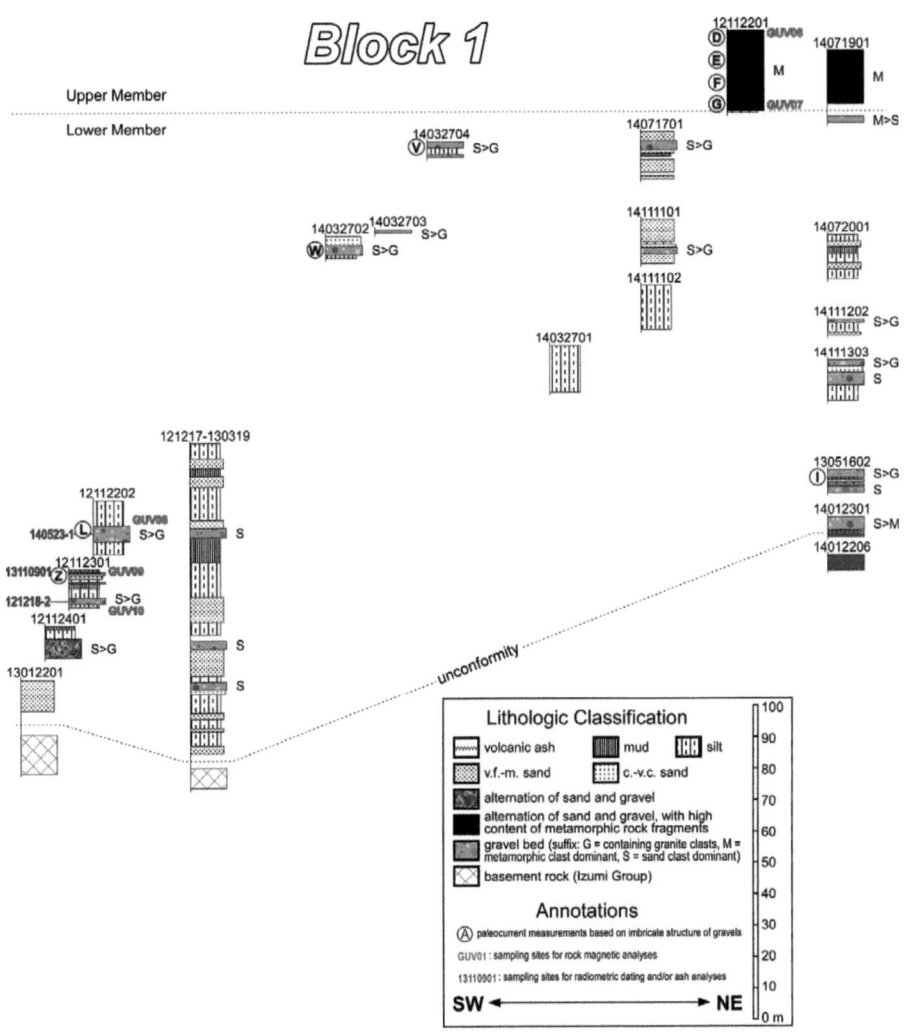

Figure 7a. Columnar sections: Block 1 (see Figure 6 for locations).

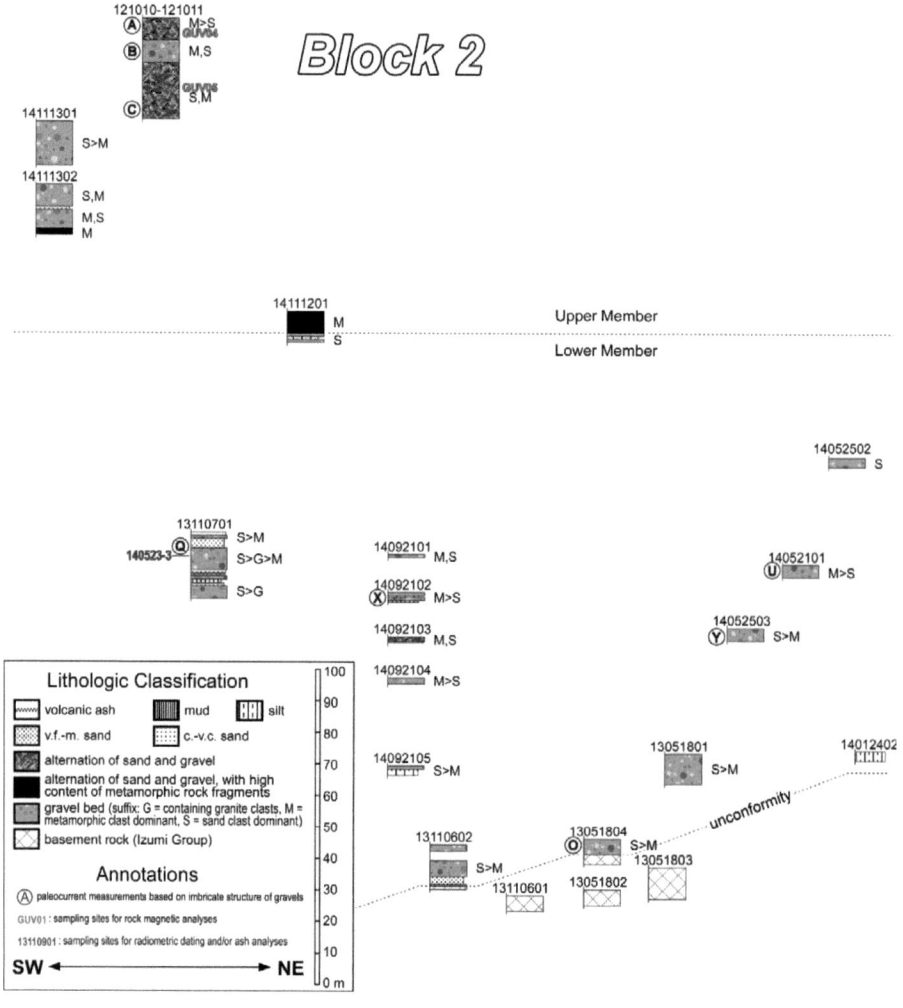

Figure 7b. Columnar sections: Block 2 (see Figure 6 for locations).

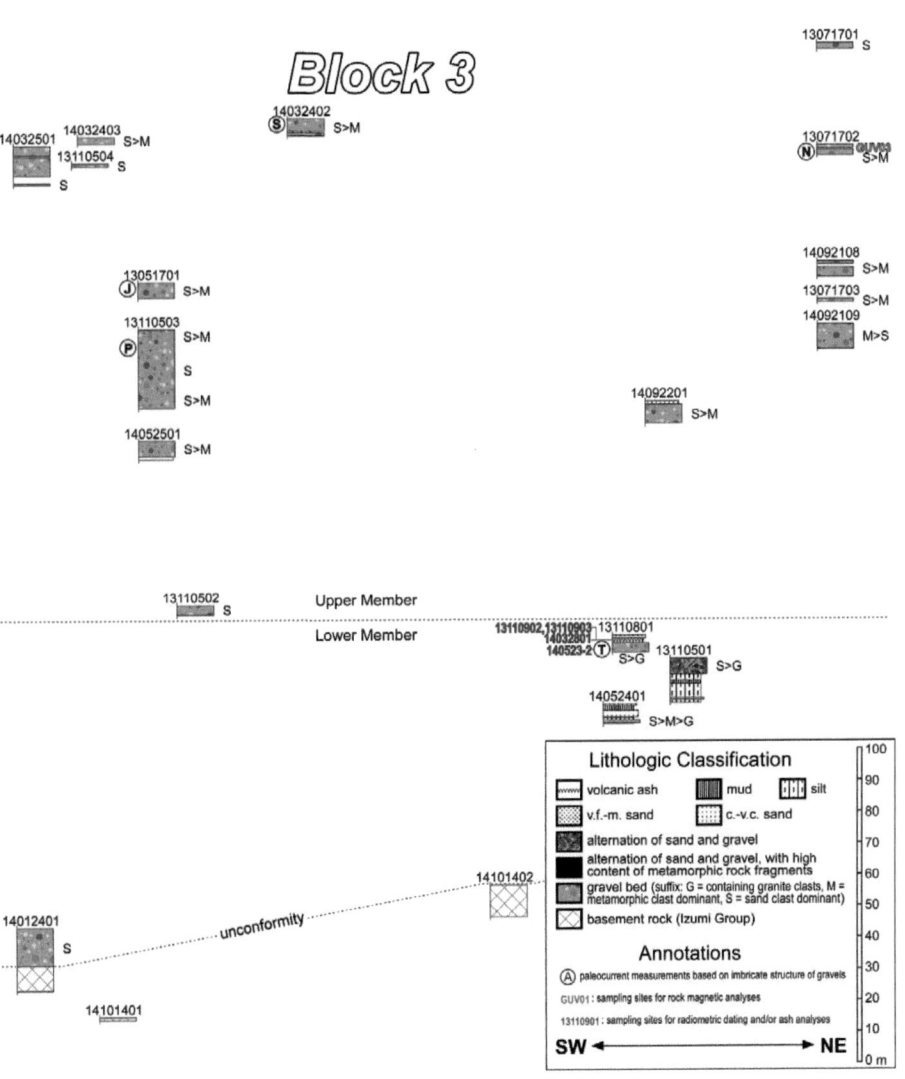

Figure 7c. Columnar sections: Block 3 (see Figure 6 for locations).

Block 4

 14032401 S

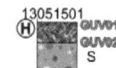 13051501 GUV01 GUV02 S

13051601

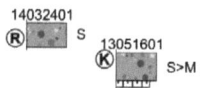 S>M

13071704 M>S

Upper Member

- -

Lower Member

13072001 S>M>G

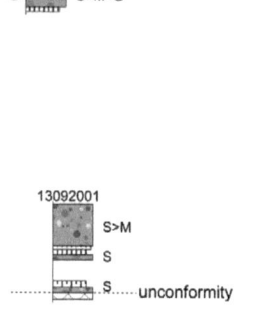

13092001 S>M S S unconformity

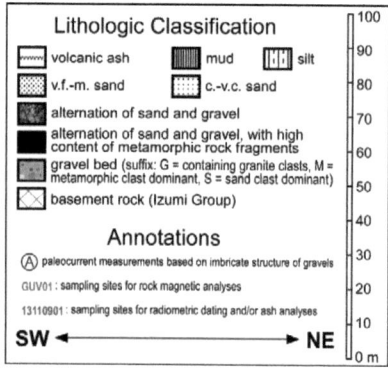

Lithologic Classification

volcanic ash	mud	silt
v.f.-m. sand	c.-v.c. sand	

alternation of sand and gravel

alternation of sand and gravel, with high content of metamorphic rock fragments

gravel bed (suffix: G = containing granite clasts, M = metamorphic clast dominant, S = sand clast dominant)

basement rock (Izumi Group)

Annotations

(A) paleocurrent measurements based on imbricate structure of gravels

GUV01 : sampling sites for rock magnetic analyses

13110901 : sampling sites for radiometric dating and/or ash analyses

SW ◄──────────────────► NE

100
90
80
70
60
50
40
30
20
10
0 m

Figure 7d. Columnar sections: Block 4 (see Figure 6 for locations).

3.1. Stratigraphic units

The Lower Member of the Gunchu Formation consists of alternating beds of partly coaly mud, silt, sand and gravel. The gravel beds' contents vary laterally. For example, a continuous section along a stream in the southern part of the distribution (Loc. 121217-130319 in Figure 7a) is composed of muddy fine sediments intercalated by thin gravel beds containing rounded sandstone pebbles. On the other hand, outcrops in the middle to northern part of the distribution (e.g., Locs. 13110701, 13110602 and 14052101 in Figure 7b, Loc. 14012401 in Figure 7c and Loc. 13092001 in Figure 7d) are thick gravel beds containing polymictic granules to boulders of varying roundness. The total thickness of the unit is about 250 m in the southern part and 180 m in the north.

Two conspicuous ash layers were identified in the Lower Member during the survey. Near the base of the unit (Loc. 12112301 in Figure 7a), bedded fine glassy ash 30 cm thick is intercalated at the top of a mud-prone sequence. A gravel bed near the top of the Lower Member (Loc. 13110801 in Figure 7c) is overlain by an upward-fining sandy-silty weathered ash 150 cm thick. Their radiometric dates are described later in the paper.

The boundary of the Lower and Upper Members is marked by an abrupt increase in the volume of the gravel beds and the emergence of monomictic clasts that originated exclusively from metamorphic rocks. For example, sandy alternations at the top of the Lower Member yielding abundant plant remains (Loc. 14071701 in Figure 7a) are overlain by gravel beds containing only schist clasts (Loc. 12112201 in Figure 7a). The contents of the schist gravels change spatially and temporally, as shown by northwardly (e.g., Loc. 14092201 in Figure 7c) and upwardly (e.g., Loc. 13051501 in Figure 7d) progressive dominance of sandstone gravels. The middle part of the Upper Member is intercalated with considerable sand/silt layers containing plant fragments (Loc. 121010-121011 in Figure 7b). Although the absence of exposure hinders estimates of the total thickness of the unit, its maximum is about 270 m.

3.2. Contact between the Gunchu Formation and the Cretaceous basement

The author found five outcrops that showed unconformable relationships between the Gunchu Formation and the Cretaceous Izumi Group. From south to north, Loc. 14012301 (Figure 7a) is composed of blocky sandstone, the undulating surface of which is covered by massive silt and gravel beds; Loc. 13110602 (Figure 7b) shows contact between hard mudstone and poorly sorted sand containing angular pebbles; Loc. 13051804 (Figure 7b) consists of angular pebbles to boulders resting on a weathered undulate crust of sandstone; Loc. 14012401 (Figure 7c) is a 40 m wide outcrop showing a sequential change in a gravel assemblage on the unconformity's surface; and an upward-fining clastic sequence covering sandstone/mudstone was confirmed at Loc. 13092001 (Figure 7d).

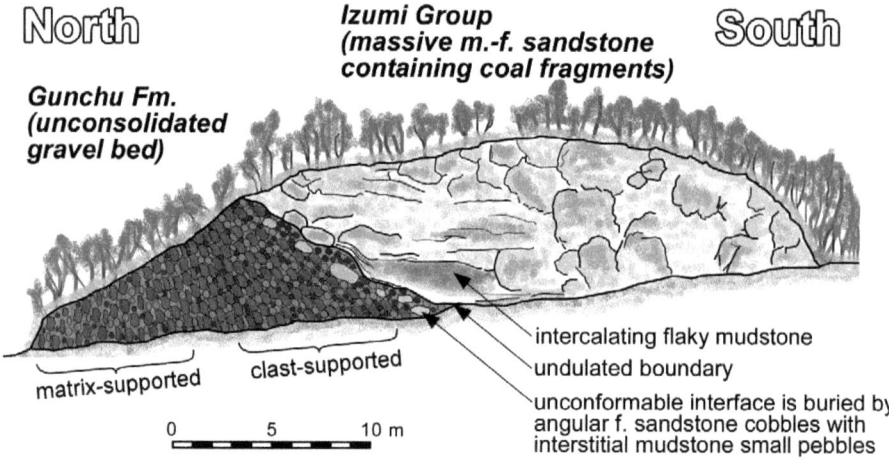

Figure 8. Sketch of the unconformity between the Cretaceous Izumi Group and the Gunchu Formation (Loc. 14012401; see Figure 6 for location).

Among the sites listed above, Loc. 14012401, in the northern part, is the most informative for understanding the nature of the unconformity. Figure 8 presents a sketch of the outcrop. On the unconformable interface, the density and size of the gravel decreases upward. It is also noteworthy that the bedding attitude in the basal part overturns so that it gradually changes upward into a vertical dip. As argued in the next section, this outcrop is adjacent to one of the Gunchu Formation's partitioning faults, and the change in dip angle may be attributed to a buildup of the intra-basin structure during its burial.

3.3. Structural trend and identification of fault blocks

The Gunchu Formation generally has a northeast strike and the Lower Member shows homoclinal, steep (~70°) dip angles toward the northwest. The northwesterly dip of the Upper Member tends to decrease (to ~30°) upward, suggesting that a tectonic event commenced during the deposition of the unit. As the strata underwent a tilting motion, concurrent activity on the faults across the sedimentary basin (broken lines in Figure 6) is inferred from offsets of the geologic boundaries. The Gunchu Formation is thought to be divided into four blocks, and in each fault block, the eroded top of the Izumi Group developed a southwest tilt, suggested by the onlapping termination pattern of the Lower Member of the Gunchu Formation onto the surface, shown by the inclined Gunchu/Izumi boundary in Figures 7a to 7c. Thus the Gunchu basin seems to have suffered syn-sedimentary deformation, the mechanism of which is considered in the last part of this book.

3.4. Sedimentology

In order to understand the provenance of the voluminous clastics filling the Pleistocene basin, the author executed sedimentological studies of the gravel beds of the Gunchu Formation. Figures 9a to 9d show 26 sites where compositional data were collected in the course of measuring the imbricated structures of the gravels. Note that granite is not grouped alone in the pie charts, since its fragile nature and spherical shape may cause underestimation of the contents during the imbrication measurements.

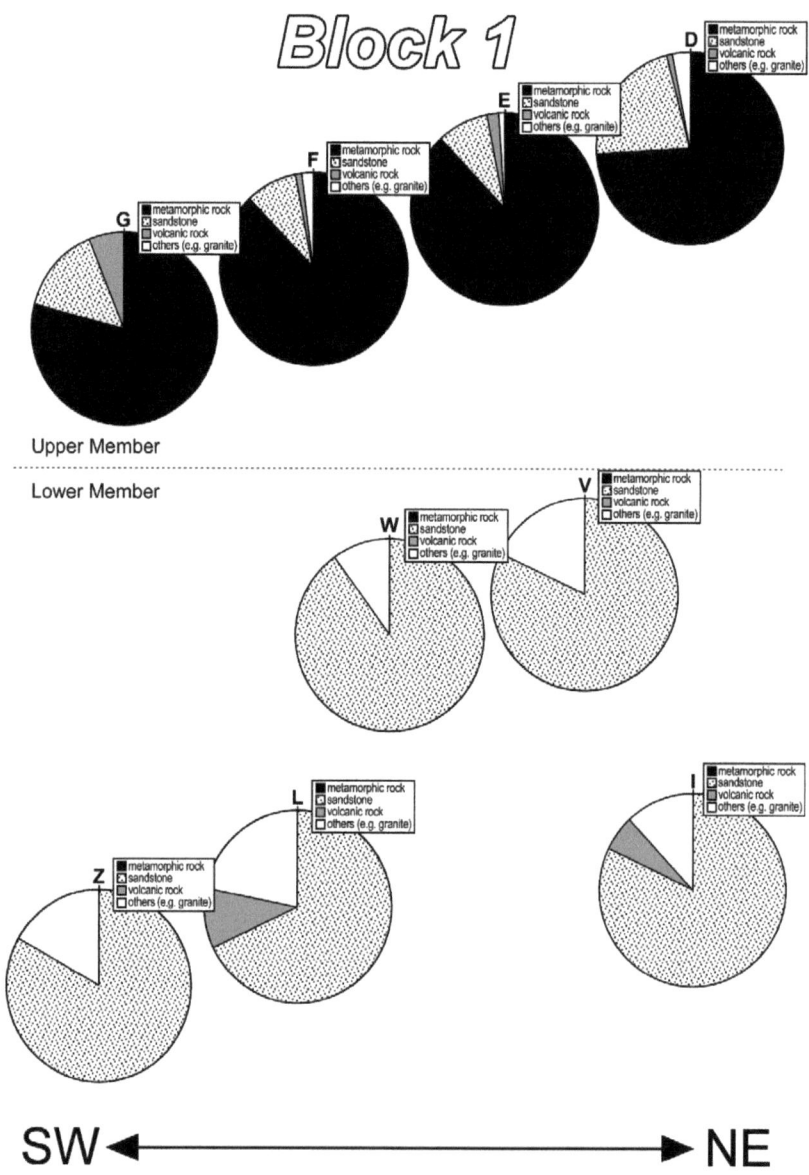

Figure 9a. Gravel composition of selected horizons. There are 100 data points for each measured site. Analysis was done for oblate pebbles, for which imbricate structures were determined. See Figure 6 for locations: Block 1.

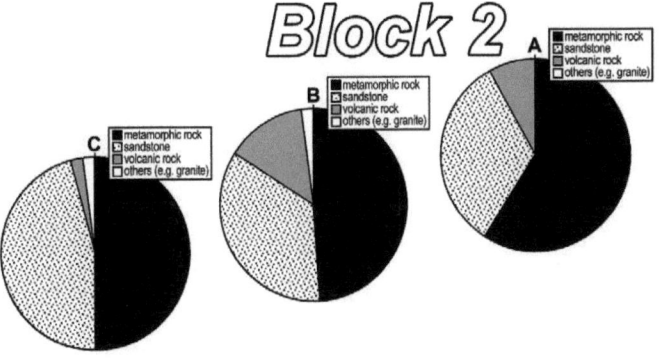

Upper Member

Lower Member

SW ◀━━━━━━━━━━━━━━━━━━━━━━▶ NE

Figure 9b. Gravel composition of selected horizons. There are 100 data points for each measured site. Analysis was done for oblate pebbles, for which imbricate structures were determined. See Figure 6 for locations: Block 2.

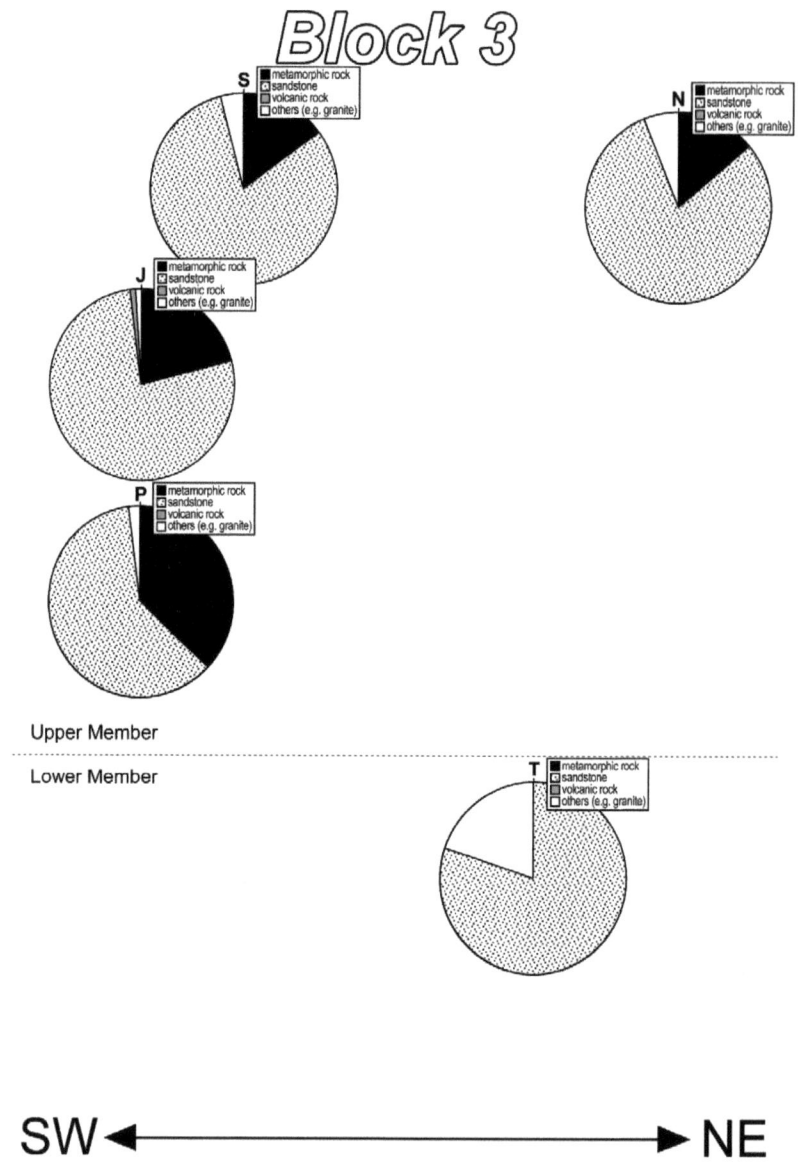

Figure 9c. Gravel composition of selected horizons. There are 100 data points for each measured site. Analysis was done for oblate pebbles, for which imbricate structures were determined. See Figure 6 for locations: Block 3.

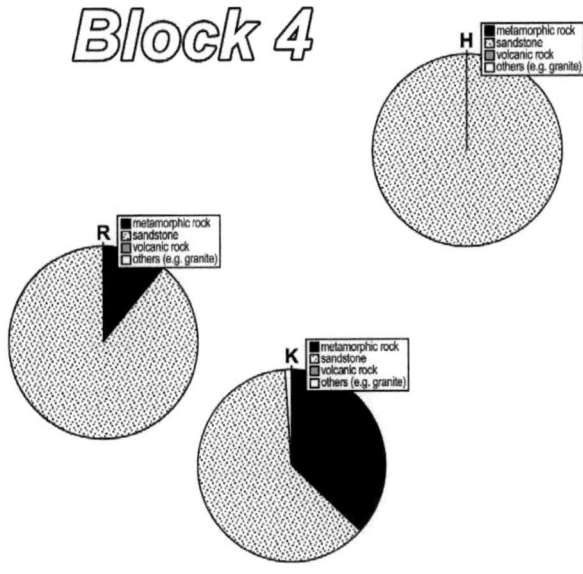

Upper Member

- -

Lower Member

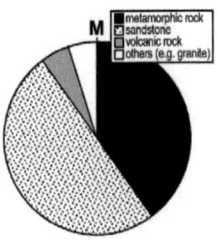

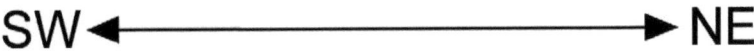

SW ◄—————————————► NE

Figure 9d. Gravel composition of selected horizons. There are 100 data points for each measured site. Analysis was done for oblate pebbles, for which imbricate structures were determined. See Figure 6 for locations: Block 4.

Obviously the composition of the gravels fluctuates spatially and temporally. Changes in the lithofacies are also shown in photographs of typical outcrops in Figures 10a and 10b. The Lower Member contains considerable amount of granite, which is categorized as "others" in the charts. Some of the measured gravel beds have matrices made of decomposed granite soil, such as point Q in Figure 9b (obtained from Loc. 13110701 in Figure 7b) and point T in Figure 9c (obtained from Loc. 13110801 in Figure 7c). Thus, the actual concentration of granitic material tends to be higher than the statistics.

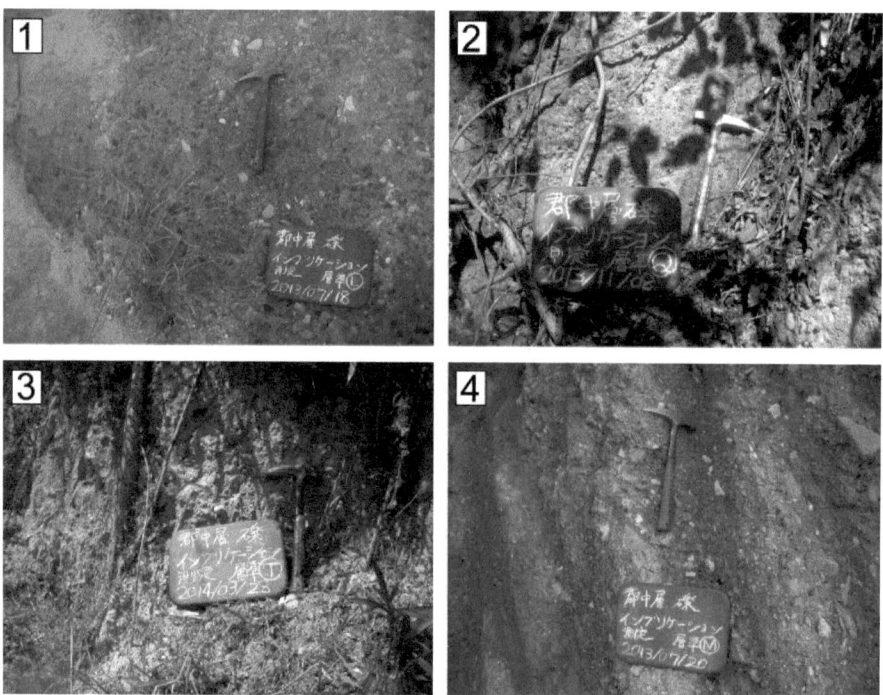

Figure 10a. Photos showing typical lithofacies of the Lower Member of the Gunchu Formation. 1: Gravel bed containing granite/chert clasts and massive silt yielding abundant plant fragments (Loc. 12112202 in Block 1). 2: Gravel bed in which matrix is rich in broken-down granite particles (Loc. 13110701 in Block 2). 3: Gravel bed containing abundant granitic material (Loc. 13110801 in Block 3). 4: Gravel bed intercalated by pebbly silt (Loc. 13072001 in Block 4). See Figure 6 for locations.

Figure 10b. Photos showing typical lithofacies of the Upper Member of the Gunchu Formation. 1: Alternation of gravel beds with high content of schist clasts and greenish gray pebbly silt rich in broken-down metamorphic particles (Loc. 12112201 in Block 1). 2: Alternation of gravel beds and sandy silt containing plant fragments (Loc. 121010-121011 in Block 2). 3: Angular gravel bed intercalated with lenticular poorly sorted sand (Loc. 13071702 in Block 3). 4: Gravel bed containing only sandstone clasts (Loc. 13051501 in Block 4). See Figure 6 for locations.

The compositional contrast between the Lower and Upper Members is most remarkable in Block 1 (Figure 9a) where an extremely high content of schist gravels characterize the Upper Member. Generally speaking, a wax and wane in the supply of metamorphic clasts is suggested through the depositional period of the Gunchu Formation.

As shown in the geologic map in Figure 3, the granite and schist clasts were not derived from terranes adjoining the sedimentary basin. The mechanism of spatiotemporal changes in the sediment provenance of the Gunchu Formation is discussed later, integrating all the other geologic information.

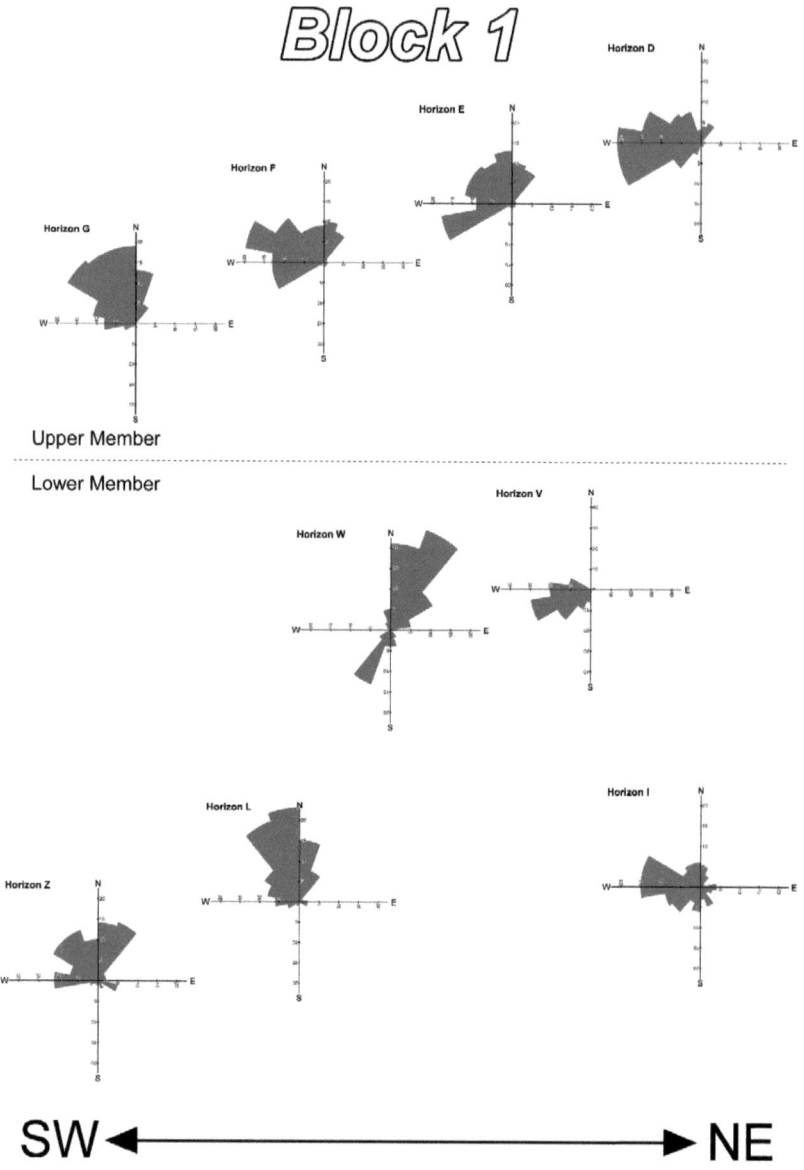

Figure 11a. Rose diagrams showing paleocurrent directions (downcurrent) of selected horizons measured using the imbricate structure of the gravels. There are 100 data points for each measured site. See Figure 6 for locations: Block 1.

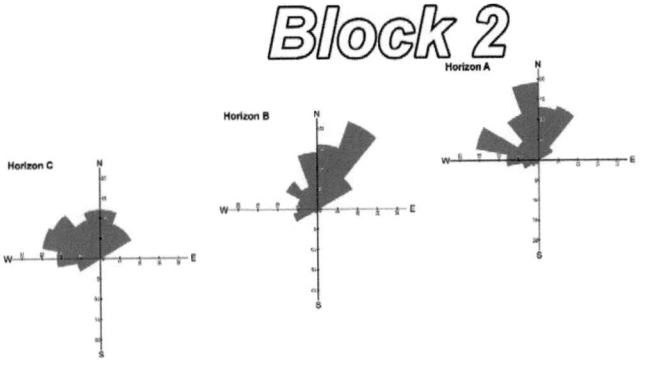

Upper Member

....

Lower Member

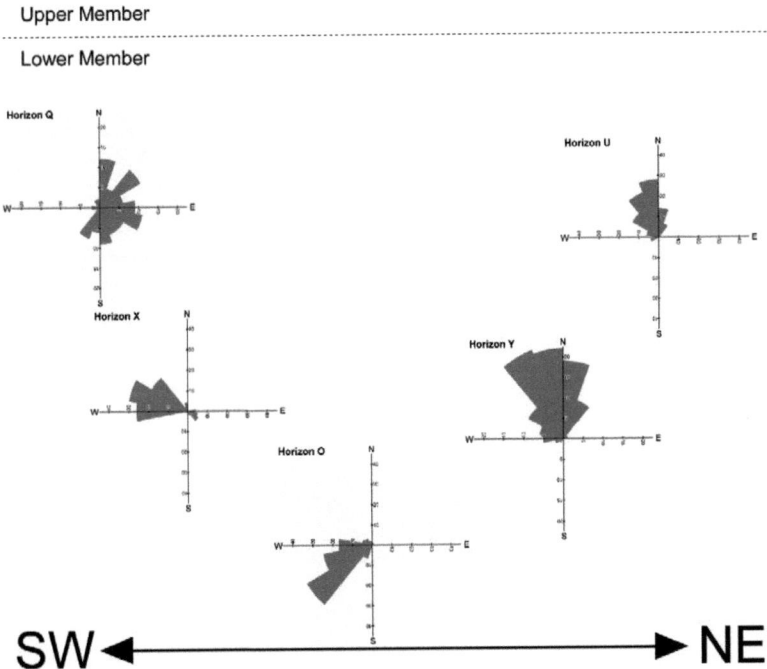

SW ◄─────────────────────────► NE

Figure 11b. Rose diagrams showing paleocurrent directions (downcurrent) of selected horizons measured using the imbricate structure of the gravels. There are 100 data points for each measured site. See Figure 6 for locations: Block 2.

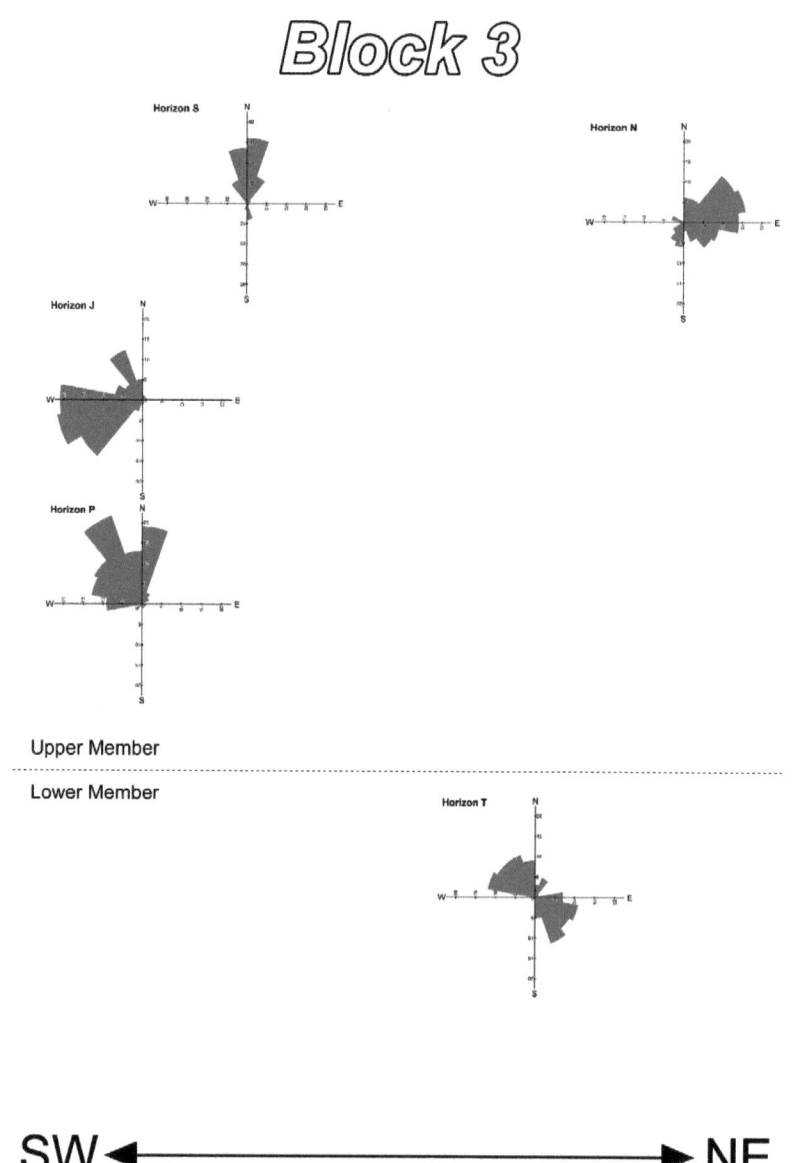

Figure 11c. Rose diagrams showing paleocurrent directions (downcurrent) of selected horizons measured using the imbricate structure of the gravels. There are 100 data points for each measured site. See Figure 6 for locations: Block 3.

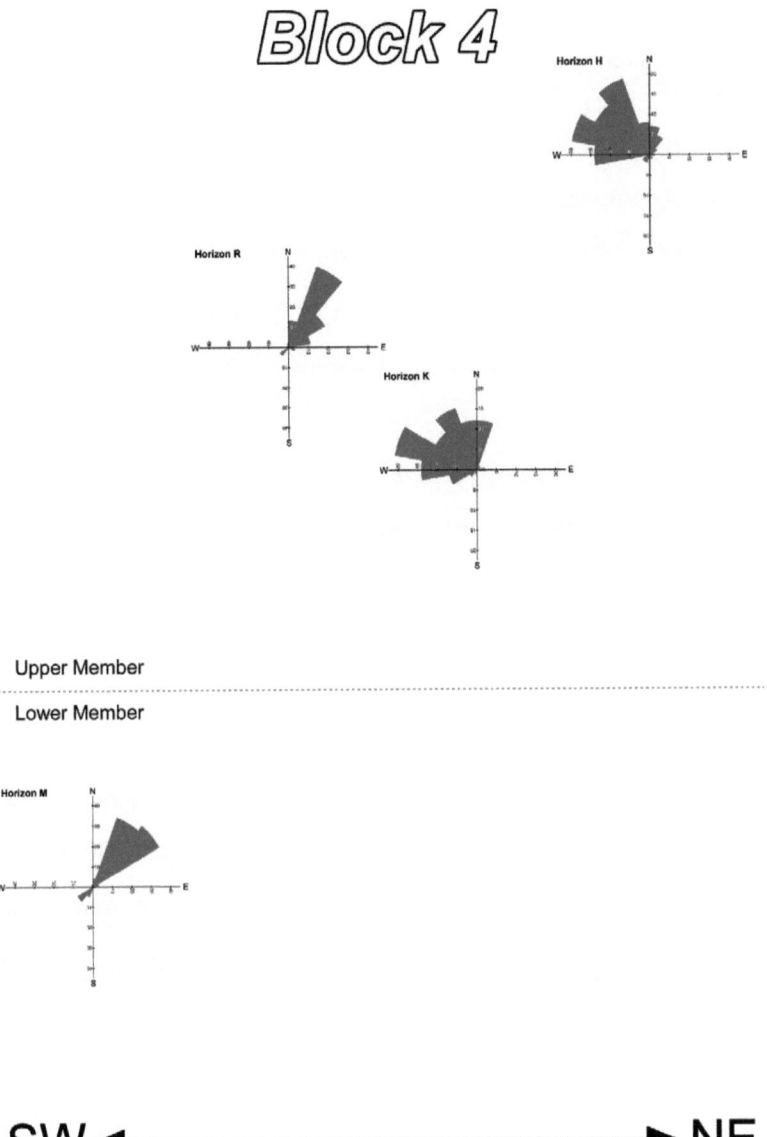

Figure 11d. Rose diagrams showing paleocurrent directions (downcurrent) of selected horizons measured using the imbricate structure of the gravels. There are 100 data points for each measured site. See Figure 6 for locations: Block 4.

Figures 11a to 11d present the paleocurrent direction for 26 sites based on the imbrication structure of their gravels, which also shows considerable variation. It seems that the NE-SW elongate Gunchu basin was alternately buried by influxes along the basin axis from inlets on the southwest (e.g., Horizon W in Figure 11a, Horizon B in Figure 11b, Horizon N in Figure 11c and Horizon R in Figure 11d) and the northeast (e.g., Horizon V in Figure 11a, Horizon O in Figure 11b and Horizon J in Figure 11c), and from uplifted areas on the southeastern side of the basin (including metamorphosed terranes in the Outer Zone; e.g., Horizon G in Figure 11a, Horizon U in Figure 11b, Horizon P in Figure 11c and Horizon H in Figure 11d).

4. Tephra analysis

4.1. Samples

Three ash layers were sampled from the Lower Member of the Gunchu Formation. Sample 13110901, obtained from a coastal outcrop (Loc. 12112301 in Block 1; Figure 6) at the basal part of the Lower Member, is a 30 cm thick light gray glassy fine ash showing wavy bedding. Samples 13110902 and 13110903 were obtained from an inland outcrop (Loc. 13110801 in Block 3; Figure 6) near the top of the Lower Member. Sample 13110902 is a 50 cm thick white fine ash underlain by a 100 cm thick brownish gray laminated sandy ash (Sp. 13110903) that contains abundant mica particles and granules.

4.2. Methods

After 30 g of the samples were weighed out, they were dried in an oven at 50°C for 15 hours. The dried samples were ultrasonically cleaned and washed in water. We wet-sieved (60, 120 and 250 mesh sieves) the samples, dried them, and then weighed each size fraction to determine the grain size distribution. Then we performed a modal analysis of the very fine fraction. We counted 200 grains and categorized them into volcanic glass, light minerals, heavy minerals, rock fragments and others, then, made microscopic observations to identify heavy minerals and the types of glass present.

4.3. Results

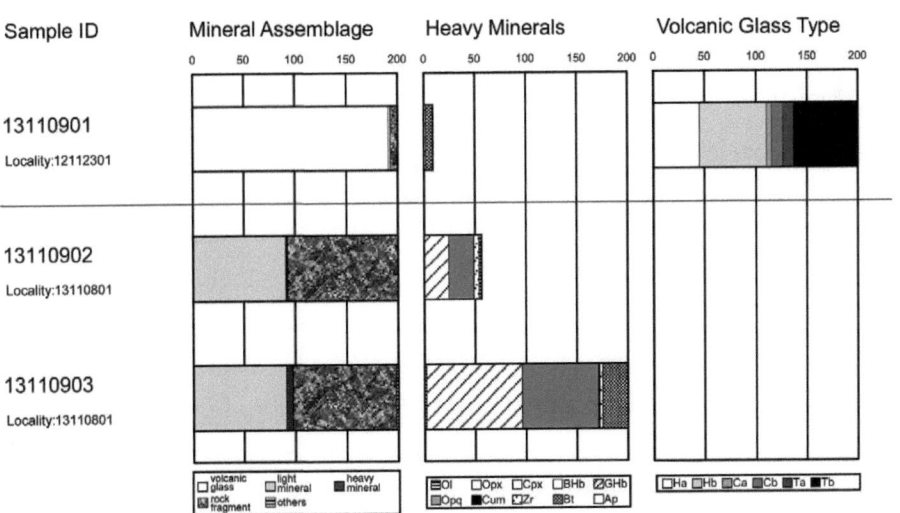

Figure 12. Results of tephra analyses. Abbreviations: Ol, olivine; Opx, orthopyroxene; Cpx, clinopyroxene; BHb, brown hornblende; GHb, green hornblende; Opq, opaque minerals; Cum, cummingtonite; Zr, zircon; Bt, biotite; Ap, apatite; Ha/Hb, bubble-wall type; Ca/Cb, intermediate type; Ta/Tb, pumice type. Classification of glass is after Yoshikawa (1976). See Figure 6 for locations.

Figure 12 presents the analytical results for the ash layers intercalated in the Lower Member of the Gunchu Formation. Sample 13110901 (Loc. 12112301), from a coastal outcrop, is a typical acidic glassy ash containing both bubble-wall and pumice-type glasses. It also contains small amounts of biotite, zircon and opaque minerals. Mizuno (1987) reported three ash layers in the basal part of the Lower Member in the coastal section, but their lithologic characteristics and mineral assemblages do not match the present ash analysis. Since Mizuno did not give detailed locations in his report, the ashes cannot be correlated in this case.

Sample 13110902 (Loc. 13110801) contains no distinguishable glass shards, but does contain small amounts of green hornblende, zircon and opaque minerals. Sample 13110903 (Loc. 13110801) also has no distinguishable glass shards, but does have common green hornblende, biotite and opaque minerals, and rare zircon, orthopyroxene, cummingtonite and epidote. Although these two samples (13110902 and 13110903), obtained from a hillside site, are lacking in visible glasses, they are accompanied by abundant glassy rock fragments and regarded as minor volcanic ash.

5. Radiometric dating

5.1. Tephra: Samples

In order to fix the stratigraphic position of the Gunchu Formation, we double-dated the zircons using the fission-track (FT) and U-Pb methods for an ash layer near the top of the Lower Member of the Gunchu Formation. Sample 14032801 was taken from the Sp. 13110902 and Sp. 13110903 horizons for tephra analysis at Loc. 13110801 in Block 3 (Figure 6). It consists of a lower brownish gray laminated sandy ash with abundant mica particles and granules 100 cm thick and an upper white fine ash 50 cm thick.

5.2. Tephra: Methods

FT dating method is after Danhara and Iwano (2009). U-Pb age data were obtained using inductively coupled plasma-mass spectrometry (ICP-MS) combined with an excimer laser ablation (LA) sample introduction system. The U-Pb age determinations on zircon samples were performed after chemical leaching using 47% HF for 20 hours at room temperature, or after FT etching using a KOH-NaOH eutectic solution for 40 hours at 225°C.

5.3. Tephra: Results

Table 1a. Fission-track grain ages of zircons obtained from Sp. 14032801.

No.	Ns	Nu	S x10^{-5} (cm^2)	ρ_s x10^{-7} (cm^{-2})	ρ_u x10^{-9} (cm^{-2})	Ns / Nu x10^3	T (Ma)	σ_T (Ma)	U (ppm)
17	42	7851	2.70	1.56	2.91	5.35	20.23	3.21	244
16	9	1727	2.40	0.38	0.72	5.21	19.71	6.63	60
15	14	4057	2.00	0.70	2.03	3.45	13.06	3.53	170
9	24	7544	3.60	0.67	2.10	3.18	12.04	2.50	176
18	27	8862	3.60	0.75	2.46	3.05	11.53	2.26	207
14	6	3416	1.80	0.33	1.90	1.76	6.65	2.73	159
29	5	2925	1.40	0.36	2.09	1.71	6.47	2.91	175
20	6	6586	4.00	0.15	1.65	0.91	3.45	1.41	138
25	14	15860	3.00	0.47	5.29	0.88	3.34	0.90	444
7	12	15490	3.20	0.38	4.84	0.77	2.93	0.85	406
28	5	7567	3.60	0.14	2.10	0.66	2.50	1.12	176
21	8	12181	2.50	0.32	4.87	0.66	2.49	0.88	409
19	6	9506	1.50	0.40	6.34	0.63	2.39	0.98	532
13	8	12943	3.20	0.25	4.04	0.62	2.34	0.83	340
8	8	14599	2.40	0.33	6.08	0.55	2.08	0.74	511
23	5	9630	2.40	0.21	4.01	0.52	1.97	0.88	337
4	12	23121	3.00	0.40	7.71	0.52	1.97	0.57	647
24	9	19949	2.70	0.33	7.39	0.45	1.71	0.57	620
2	5	11146	2.50	0.20	4.46	0.45	1.70	0.76	374
5	7	16968	3.60	0.19	4.71	0.41	1.56	0.59	396
22	12	29427	4.00	0.30	7.36	0.41	1.54	0.45	618
27	6	14732	1.80	0.33	8.18	0.41	1.54	0.63	687
6	6	15363	3.60	0.17	4.27	0.39	1.48	0.61	358
3	10	25796	5.00	0.20	5.16	0.39	1.47	0.47	433
1	4	11236	3.60	0.11	3.12	0.36	1.35	0.68	262
12	5	14140	2.40	0.21	5.89	0.35	1.34	0.60	495
11	4	12153	2.40	0.17	5.06	0.33	1.25	0.63	425
10	2	6535	1.80	0.11	3.63	0.31	1.16	0.82	305
26	6	25095	4.00	0.15	6.27	0.24	0.91	0.37	527
30	1	4486	1.50	0.07	2.99	0.22	0.84	0.84	251

Ns is number of spontaneous tracks, Nu is number of ^{238}U counts, S is analyzed area of crystal, ρ_s is density of spontaneous tracks, ρ_u is density of ^{238}U counts, σ_T is error for each grain age (1σ), U is uranium density. Uranium concentration for standardization using 91500 standard zircon is 1.124x10^6 cm^2. Epsilon (ϵ) corresponding to conventional zeta (ζ) value for Fish Canyon Tuff zircon is 33.7±1.2.

Results of FT dating for 30 zircon grains are summarized in Table 1a. As shown in Figure 13a, we excluded seven grains (their numbers annotated in the histogram correspond to sample numbers in Table 1a) having significantly old ages as reworked crystals, and produced positive result in a chi-squared test. The average age of the selected grains is 1.8±0.2 Ma (error: 1σ). Figure 13b indicates that the U-Pb ages for Sp. 14032801 basically concur with the FT ages for the same sample.

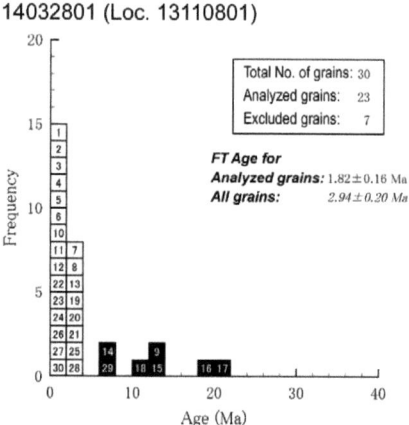

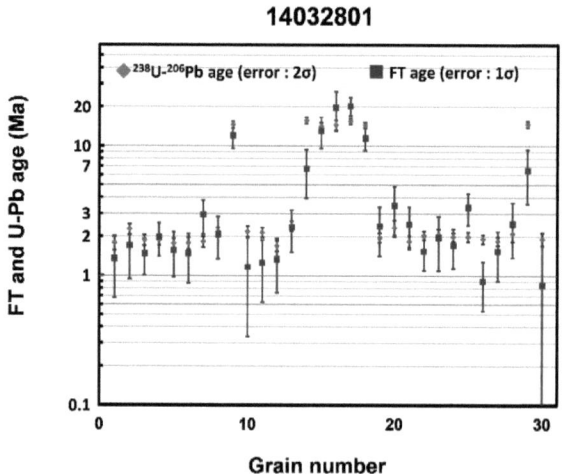

Figure 13a. Result of fission-track (FT) dating of zircons obtained from a tephra intercalated in the Gunchu Formation. See Figure 6 for location (Loc. 13110801).

Figure 13b. FT and U-Pb age plots for zircons obtained from a tephra intercalated in the Gunchu Formation. See Figure 6 for location (Loc. 13110801).

Table 1b. U-Pb ages for all analyzed zircon grains obtained from Sp. 14032801.

sample name	Total count 206Pb	208Pb	232Th	235U	238U	Th/U	Isotopic ratios 208Pb/206Pb	Error 2σ	206Pb/238U	Error 2σ	207Pb/235U	Error 2σ	Disc.*	Age (Ma) 206Pb/238U	Error 2σ	207Pb/235U	Error 2σ
14032801 no.1	294	54	262994	7106	973994	0.61	0.1837	± 0.0058	0.00028	± 0.000033	0.0070	± 0.0019	-172	1.79	± 0.22	7.06	± 1.97
14032801 no.2	535	28	311994	10154	1399994	0.51	0.0523	± 0.0016	0.00035	± 0.000032	0.0025	± 0.0010	concordant	2.28	± 0.21	2.57	± 0.98
14032801 no.3	508	29	379994	11749	1619994	0.53	0.0671	± 0.0018	0.00029	± 0.000027	0.0023	± 0.0009	concordant	1.87	± 0.18	2.30	± 0.86
14032801 no.4	759	24	589994	17551	2419994	0.55	0.0316	± 0.0010	0.00029	± 0.000023	0.0013	± 0.0005	concordant	1.87	± 0.15	1.27	± 0.53
14032801 no.5	438	21	245994	10734	1479994	0.38	0.0479	± 0.0015	0.00027	± 0.000027	0.0018	± 0.0006	concordant	1.76	± 0.18	1.82	± 0.80
14032801 no.6	400	32	323994	9719	1339994	0.55	0.0600	± 0.0025	0.00029	± 0.000029	0.0030	± 0.0011	concordant	1.78	± 0.19	3.06	± 1.10
14032801 no.7	466	71	361994	11024	1519994	0.54	0.1524	± 0.0048	0.00028	± 0.000028	0.0059	± 0.0014	-138	1.83	± 0.18	5.99	± 1.47
14032801 no.8	663	27	401994	13863	1909994	0.48	0.0390	± 0.0012	0.00034	± 0.000028	0.0018	± 0.0007	concordant	2.16	± 0.18	1.82	± 0.71
14032801 no.9	1611	94	104994	4772	657994	0.36	0.0583	± 0.0018	0.00226	± 0.000133	0.0181	± 0.0039	-138	14.58	± 0.86	18.20	± 3.95
14032801 no.10	416	39	168994	8268	1139994	0.36	0.0938	± 0.0029	0.00034	± 0.000035	0.0043	± 0.0014	-26	2.18	± 0.22	4.39	± 1.43
14032801 no.11	572	59	433994	11532	1589994	0.62	0.1031	± 0.0032	0.00033	± 0.000030	0.0047	± 0.0013	-54	2.14	± 0.19	4.76	± 1.27
14032801 no.12	525	23	448994	13417	1849994	0.55	0.0438	± 0.0014	0.00026	± 0.000024	0.0016	± 0.0007	concordant	1.69	± 0.16	1.60	± 0.67
14032801 no.13	505	15	274992	9211	1269989	0.51	0.0287	± 0.0013	0.00037	± 0.000034	0.0015	± 0.0008	concordant	2.41	± 0.22	1.47	± 0.78
14032801 no.14	1558	84	113992	4323	595989	0.45	0.0536	± 0.0024	0.00245	± 0.000135	0.0178	± 0.0040	concordant	15.80	± 0.87	17.94	± 4.07
14032801 no.15	1508	75	68992	4620	636989	0.26	0.0494	± 0.0023	0.00222	± 0.000124	0.0149	± 0.0035	concordant	14.31	± 0.80	15.00	± 3.59
14032801 no.16	544	20	40192	1639	225989	0.42	0.0358	± 0.0016	0.00226	± 0.000200	0.0110	± 0.0050	concordant	14.55	± 1.29	11.09	± 5.09
14032801 no.17	2348	137	110892	6622	912989	0.29	0.0681	± 0.0027	0.00241	± 0.000112	0.0190	± 0.0034	concordant	15.54	± 0.72	19.13	± 3.45
14032801 no.18	1848	68	274992	5606	772989	0.84	0.0365	± 0.0017	0.00224	± 0.000114	0.0111	± 0.0028	concordant	14.45	± 0.74	11.22	± 2.81
14032801 no.19	644	64	555992	14433	1988989	0.66	0.0996	± 0.0045	0.00030	± 0.000025	0.0041	± 0.0010	-48	1.96	± 0.16	4.11	± 1.05
14032801 no.20	200	6	47092	3750	516989	0.22	0.0275	± 0.0013	0.00036	± 0.000036	0.0014	± 0.0012	concordant	2.34	± 0.33	1.37	± 1.17
14032801 no.21	465	20	361992	11097	1529989	0.56	0.0419	± 0.0019	0.00029	± 0.000032	0.0016	± 0.0007	concordant	1.84	± 0.17	1.65	± 0.75
14032801 no.22	779	34	626992	16754	2309989	0.64	0.0430	± 0.0020	0.00032	± 0.000024	0.0018	± 0.0006	concordant	2.04	± 0.15	1.87	± 0.65
14032801 no.23	434	40	189992	9138	1259989	0.36	0.0910	± 0.0042	0.00032	± 0.000023	0.0040	± 0.0013	-22	2.08	± 0.20	4.04	± 1.30
14032801 no.24	764	50	681992	16826	2319989	0.70	0.0648	± 0.0019	0.00031	± 0.000027	0.0027	± 0.0008	concordant	1.99	± 0.15	2.75	± 0.79
14032801 no.25	546	29	448992	12039	1659989	0.64	0.0522	± 0.0025	0.00031	± 0.000025	0.0022	± 0.0008	concordant	1.99	± 0.18	2.22	± 0.84
14032801 no.26	625	27	486992	14288	1969989	0.59	0.0424	± 0.0019	0.00030	± 0.000030	0.0017	± 0.0007	concordant	1.92	± 0.16	1.74	± 0.68
14032801 no.27	785	33	518992	16639	2569989	0.48	0.0414	± 0.0019	0.00029	± 0.000021	0.0016	± 0.0006	concordant	1.85	± 0.14	1.63	± 0.58
14032801 no.28	230	4	161992	4787	659989	0.58	-0.0152	± -0.0007	0.00033	± 0.000044	-0.0007	± 0.0006	concordant	2.11	± 0.28	-0.69	± 3.58
14032801 no.29	1598	79	92392	4758	655989	0.33	0.0491	± 0.0022	0.00229	± 0.000124	0.0152	± 0.0035	concordant	14.72	± 0.80	15.34	± 3.58
14032801 no.30	298	4	129992	6810	938989	0.33	0.0117	± 0.0005	0.00030	± 0.000035	0.0005	± 0.0005	145	1.92	± 0.23	0.48	± 0.52
Average ± 2SD														4.99	± 11.08	5.57	± 11.85
Weighted average														2.14	± 0.04	2.16	± 0.17

Th/U is ratio of thorium and uranium concentrations. Disc.* is degree of discordance. Positive and negative values mean left- and right-side significant offsets from concordia line, respectively.

Table 1c: U-Pb ages for adopted zircon grains obtained from Sp. 14032801.

sample name	Total count					Th/U	Isotopic ratios						Disc.*	Age (Ma)			
	^{206}Pb	^{208}Pb	^{232}Th	^{235}U	^{238}U		^{208}Pb/^{206}Pb	Error 2σ	^{206}Pb/^{238}U	Error 2σ	^{207}Pb/^{235}U	Error 2σ		^{206}Pb/^{238}U	Error 2σ	^{207}Pb/^{235}U	Error 2σ
14032801 no.2	535	28	311994	10154	1369964	0.51	0.05234 ± 0.00164	0.00035 ± 0.00003	0.00253 ± 0.00097				concordant	2.28 ± 0.21		2.57 ± 0.98	
14032801 no.3	508	29	379994	11749	1619964	0.53	0.05709 ± 0.00179	0.00029 ± 0.00003	0.00227 ± 0.00085				concordant	1.87 ± 0.18		2.30 ± 0.86	
14032801 no.4	759	24	589994	17551	2419994	0.55	0.03162 ± 0.00069	0.00029 ± 0.00002	0.00126 ± 0.00052				concordant	1.87 ± 0.15		1.27 ± 0.53	
14032801 no.5	438	21	245994	10734	1479994	0.38	0.04795 ± 0.00151	0.00027 ± 0.00003	0.00180 ± 0.00079				concordant	1.76 ± 0.18		1.82 ± 0.80	
14032801 no.6	400	32	323994	9719	1339994	0.55	0.06000 ± 0.00251	0.00028 ± 0.00003	0.00302 ± 0.00108				concordant	1.78 ± 0.19		3.06 ± 1.10	
14032801 no.8	693	27	401994	13853	1909994	0.48	0.03896 ± 0.00122	0.00034 ± 0.00003	0.00179 ± 0.00070				concordant	2.16 ± 0.18		1.82 ± 0.71	
14032801 no.12	525	23	448994	13417	1849994	0.55	0.04381 ± 0.00138	0.00026 ± 0.00002	0.00157 ± 0.00066				concordant	1.69 ± 0.16		1.60 ± 0.67	
14032801 no.20	200	6	47092	3750	516989	0.22	0.02750 ± 0.00125	0.00036 ± 0.00005	0.00135 ± 0.00116				concordant	2.34 ± 0.33		1.37 ± 1.17	
14032801 no.21	465	20	361992	11097	1529989	0.56	0.04194 ± 0.00191	0.00029 ± 0.00003	0.00162 ± 0.00074				concordant	1.84 ± 0.17		1.65 ± 0.75	
14032801 no.22	779	34	626992	16754	2309989	0.64	0.04300 ± 0.00196	0.00032 ± 0.00002	0.00185 ± 0.00064				concordant	2.04 ± 0.15		1.87 ± 0.65	
14032801 no.24	764	50	681992	16826	2319989	0.70	0.06479 ± 0.00296	0.00031 ± 0.00002	0.00271 ± 0.00078				concordant	1.99 ± 0.15		2.75 ± 0.79	
14032801 no.25	546	29	448992	12039	1659989	0.64	0.05220 ± 0.00238	0.00031 ± 0.00003	0.00218 ± 0.00083				concordant	1.99 ± 0.18		2.22 ± 0.84	
14032801 no.26	625	27	486992	14288	1969989	0.59	0.04240 ± 0.00193	0.00030 ± 0.00003	0.00171 ± 0.00067				concordant	1.92 ± 0.16		1.74 ± 0.68	
14032801 no.27	785	33	518992	18639	2569989	0.48	0.04140 ± 0.00189	0.00029 ± 0.00002	0.00161 ± 0.00057				concordant	1.85 ± 0.14		1.63 ± 0.58	
													Average ± 2SD	1.96 ± 0.39		1.98 ± 1.06	
													Weighted average	1.92 ± 0.05		1.85 ± 0.20	

Th/U is ratio of thorium and uranium concentrations. Disc.* is degree of discordance. Positive and negative values mean left- and right-side significant offsets from concordia line, respectively.

38

14032801 (*Upper part of the Lower Member; Loc. 13110801*)

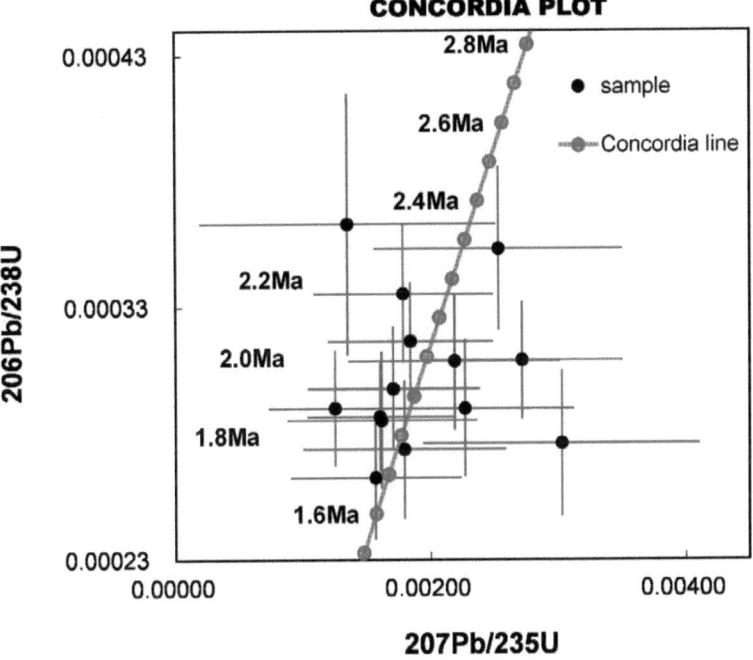

Figure 13c. Concordia plot for U-Pb ages of zircons obtained from a tephra intercalated in the Gunchu Formation. See Figure 6 for location (Loc. 13110801).

The U-Pb dating results for total 30 zircon grains are summarized in Table 1b. Based on the concordia plot for U-Pb zircon ages in Figure 13c, we excluded zircon grains with discordant ages, then adopted the younger cluster of 14 concordant grains to calculate the age. The final results are presented in Table 1c and Figure 13d. The average age of the selected grains is 1.92±0.05 Ma (error: 2σ).

As for the base of the Gunchu Formation, Sp. 13110901 for the present tephra analysis (Loc. 12112301 in Block 1; Figure 6) was dated by Kitabayashi et al. (2012). They obtained a FT age of 2.2±0.3 Ma (error: 1σ) and a U-Pb age of 2.13±0.05 Ma (error: 2σ). This confirms that the Lower Member of the Gunchu Formation was deposited within a short period in the early Pleistocene.

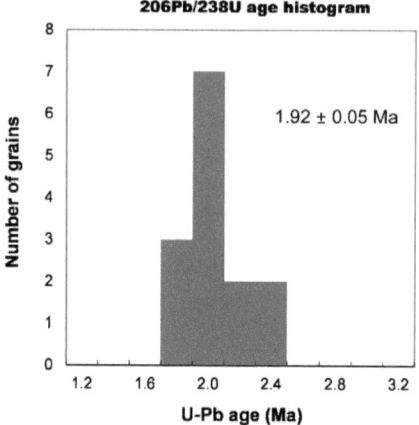

14032801 (*Upper part of the Lower Member; Loc. 13110801*)

206Pb/238U age histogram

1.92 ± 0.05 Ma

Number of grains

U-Pb age (Ma)

Figure 13d. Result of U-Pb dating of zircons obtained from a tephra intercalated in the Gunchu Formation. See Figure 6 for location (Loc. 13110801).

5.4. Granite pebble: Samples

In order to constrain the provenance of granitic clasts frequently contained in the Lower Member of the Gunchu Formation, we double-dated the zircons using the FT and U-Pb methods for four gravel beds intercalated in the Lower Member of the Gunchu Formation. In ascending order, the granite pebbles of Sp. 121218-2 (Loc. 12112301), Sp. 140523-1 (Loc. 12112202), Sp. 140523-3 (Loc. 13110701) and Sp. 140523-2 (Loc. 13110801) were collected as shown in Figure 6.

5.5. Granite pebble: Methods

The FT dating method we used is after Danhara and Iwano (2009). The U-Pb age data were obtained using inductively coupled plasma-mass spectrometry (ICP-MS) combined with an excimer laser ablation (LA) sample introduction system. The U-Pb age determinations of the zircon samples were performed after FT etching using a

KOH-NaOH eutectic solution at 225°C for 15 and 18 hours for Sp. 121218-2 and Sp. 140523-1, 2, 3, respectively.

5.6. Granite pebble: Results

In this section, we present dating results for inland sites first, followed by those for coastal sites because their clustering patterns differ slightly.

Table 2a. Fission-track grain ages of zircons obtained from Sp. 140523-2.

No.	Ns	Nu	S x10^{-5} (cm^2)	ρ_s x10^{-7} (cm^{-2})	ρ_u x10^{-9} (cm^{-2})	Ns / Nu x10^3	T (Ma)	σ_T (Ma)	U (ppm)
36	160	227	1.60	1.00	1.42	704.85	117.99	14.35	130
17	48	73	0.60	0.80	1.22	657.53	110.14	21.66	112
42	118	190	0.90	1.31	2.11	621.05	104.08	13.91	194
31	75	126	1.20	0.63	1.05	595.24	99.78	15.90	96
37	140	252	0.90	1.56	2.80	555.56	93.18	11.50	257
32	86	163	0.80	1.08	2.04	527.61	88.52	13.10	187
34	130	247	1.60	0.81	1.54	526.32	88.31	11.12	142
6	137	267	1.50	0.91	1.78	513.11	86.11	10.61	164
39	68	140	1.20	0.57	1.17	485.71	81.54	13.14	107
8	95	212	0.80	1.19	2.65	448.11	75.26	10.48	244
43	131	299	1.20	1.09	2.49	438.13	73.60	9.05	229
41	95	222	1.20	0.79	1.85	427.93	71.89	9.95	170
38	71	171	0.90	0.79	1.90	415.20	69.77	10.82	175
16	248	610	4.00	0.62	1.52	406.56	68.32	6.76	140
7	127	313	0.90	1.41	3.48	405.75	68.19	8.41	320
33	77	191	1.20	0.64	1.59	403.14	67.75	10.13	146
40	82	205	0.80	1.03	2.56	400.00	67.22	9.79	235
1	150	376	0.90	1.67	4.18	398.94	67.05	7.78	384
3	173	440	1.60	1.08	2.75	393.18	66.08	7.29	253
23	157	417	1.60	0.98	2.61	376.50	63.29	7.19	240
20	80	226	1.00	0.80	2.26	353.98	59.53	8.64	208
12	96	277	0.80	1.20	3.46	346.57	58.29	7.85	318
22	116	336	1.40	0.83	2.40	345.24	58.06	7.28	221
44	121	351	1.00	1.21	3.51	344.73	57.98	7.16	323
18	128	394	1.00	1.28	3.94	324.87	54.65	6.58	362
24	137	474	1.50	0.91	3.16	289.03	48.65	5.66	290

Ns is number of spontaneous tracks, Nu is number of ^{238}U counts, S is analyzed area of crystal, ρ_s is density of spontaneous tracks, ρ_u is density of ^{238}U counts, σ_T is error for each grain age (1σ), U is uranium density. Uranium concentration for standardization using 91500 standard zircon is 0.103x10^6 cm^2. Epsilon (ε) corresponding to conventional zeta (ζ) value for Fish Canyon Tuff zircon is 32.9±2.1.

Table 2b: U-Pb ages for all analyzed zircon grains obtained from Sp. 140523-2.

sample name	Total count					Th/U	Isotopic ratios						Disc.*	Age (Ma)			
	207Pb	206Pb	232Th	235U	238U		207Pb/206Pb	Error 2σ	206Pb/238U	Error 2σ	207Pb/235U	Error 2σ		206Pb/238U	Error 2σ	207Pb/235U	Error 2σ
2. 140523-2 no.1	3775	219	93593	2945	405994	0.25	0.0579 ± 0.0017		0.01446 ± 0.000644		0.1115 ± 0.0160		concordant	92.5 ± 4.1		107.4 ± 16.1	
2. 140523-2 no.2	5285	306	284993	3815	525994	0.59	0.0579 ± 0.0017		0.01562 ± 0.000640		0.1203 ± 0.0148		concordant	99.9 ± 4.1		115.4 ± 14.9	
2. 140523-2 no.3	2315	104	80693	1777	244994	0.36	0.0448 ± 0.0013		0.01469 ± 0.000757		0.0877 ± 0.0179		concordant	94.0 ± 4.9		85.3 ± 18.0	
2. 140523-2 no.4	6875	332	198693	5207	717994	0.30	0.0483 ± 0.0014		0.01489 ± 0.000577		0.0956 ± 0.0112		concordant	95.3 ± 3.7		92.7 ± 11.3	
2. 140523-2 no.5	2675	159	90393	2074	285994	0.35	0.0593 ± 0.0018		0.01454 ± 0.000716		0.1149 ± 0.0192		concordant	93.1 ± 4.6		110.4 ± 19.4	
2. 140523-2 no.6	884	44	33993	674	92894	0.40	0.0495 ± 0.0015		0.01480 ± 0.001095		0.0975 ± 0.0306		concordant	94.7 ± 7.1		94.5 ± 30.6	
2. 140523-2 no.7	2985	137	116693	2473	340994	0.38	0.0458 ± 0.0014		0.01361 ± 0.000648		0.0830 ± 0.0148		concordant	87.1 ± 4.2		81.0 ± 14.9	
2. 140523-2 no.8	4875	251	149693	3757	517994	0.32	0.0514 ± 0.0015		0.01463 ± 0.000611		0.1002 ± 0.0134		concordant	93.6 ± 3.9		97.0 ± 13.5	
2. 140523-2 no.9	5755	284	179993	4366	601994	0.33	0.0493 ± 0.0015		0.01488 ± 0.000697		0.0976 ± 0.0123		concordant	95.1 ± 3.8		94.5 ± 12.4	
2. 140523-2 no.10	5875	326	160993	4352	599994	0.29	0.0554 ± 0.0016		0.01522 ± 0.000609		0.1124 ± 0.0134		concordant	97.4 ± 3.9		108.1 ± 13.5	
2. 140523-2 no.11	7845	381	207993	6179	851994	0.27	0.0485 ± 0.0014		0.01432 ± 0.000541		0.0925 ± 0.0102		concordant	91.6 ± 3.5		89.8 ± 10.3	
2. 140523-2 no.12	4065	182	126993	3032	417994	0.33	0.0447 ± 0.0013		0.01512 ± 0.000660		0.0900 ± 0.0140		concordant	96.7 ± 4.3		87.5 ± 14.1	
2. 140523-2 no.13	2015	86	143993	1501	206994	0.76	0.0426 ± 0.0015		0.01513 ± 0.000817		0.0857 ± 0.0192		concordant	96.8 ± 5.3		83.5 ± 19.3	
2. 140523-2 no.14	4525	223	156993	3365	463994	0.37	0.0492 ± 0.0015		0.01516 ± 0.000644		0.0994 ± 0.0141		concordant	97.0 ± 4.2		96.2 ± 14.2	
2. 140523-2 no.15	6515	292	223993	5164	711994	0.34	0.0448 ± 0.0013		0.01423 ± 0.000577		0.0848 ± 0.0105		concordant	91.1 ± 3.6		82.7 ± 10.6	
2. 140523-2 no.16	1424	81	58692	1015	139994	0.45	0.0567 ± 0.0028		0.01575 ± 0.000977		0.1208 ± 0.0285		concordant	100.7 ± 6.3		115.8 ± 28.6	
2. 140523-2 no.17	1254	54	52892	921	126994	0.45	0.0429 ± 0.0021		0.01529 ± 0.000994		0.0886 ± 0.0252		concordant	97.8 ± 6.4		86.6 ± 25.3	
2. 140523-2 no.18	5214	279	249992	3757	517994	0.52	0.0535 ± 0.0026		0.01558 ± 0.000658		0.1127 ± 0.0150		concordant	99.7 ± 6.4		108.4 ± 15.2	
2. 140523-2 no.19	15874	843	773992	11604	1599994	0.52	0.0531 ± 0.0026		0.01536 ± 0.000546		0.1103 ± 0.0095		concordant	98.3 ± 3.5		106.2 ± 9.6	
2. 140523-2 no.20	1034	65	47192	827	1139994	0.45	0.0626 ± 0.0030		0.01404 ± 0.000684		0.1189 ± 0.0312		concordant	89.9 ± 6.3		114.1 ± 31.2	
2. 140523-2 no.21	5214	240	368992	3887	535994	0.74	0.0460 ± 0.0022		0.01506 ± 0.000636		0.0937 ± 0.0133		concordant	96.4 ± 4.1		90.9 ± 13.4	
2. 140523-2 no.22	1624	96	98392	1255	172994	0.62	0.0590 ± 0.0029		0.01453 ± 0.000859		0.1159 ± 0.0252		concordant	93.0 ± 5.5		111.3 ± 25.3	
2. 140523-2 no.23	2174	109	116992	1661	228994	0.55	0.0500 ± 0.0024		0.01470 ± 0.000787		0.0994 ± 0.0203		concordant	94.1 ± 5.1		96.3 ± 20.4	
2. 140523-2 no.24	1164	39	34392	834	114994	0.32	0.0333 ± 0.0016		0.01567 ± 0.001049		0.0706 ± 0.0234		†	100.2 ± 6.8		69.2 ± 23.5	
2. 140523-2 no.25	5794	262	235992	4460	614994	0.41	0.0452 ± 0.0022		0.01459 ± 0.000602		0.0891 ± 0.0121		concordant	93.3 ± 3.9		86.7 ± 12.3	
2. 140523-2 no.26	2904	143	36392	2205	303994	0.13	0.0492 ± 0.0024		0.01479 ± 0.000724		0.0983 ± 0.0177		concordant	94.6 ± 4.7		95.2 ± 17.8	
2. 140523-2 no.27	3974	186	187992	3104	427994	0.47	0.0467 ± 0.0023		0.01438 ± 0.000647		0.0909 ± 0.0144		concordant	92.0 ± 4.2		88.3 ± 14.6	
2. 140523-2 no.28	5634	263	207992	4359	600994	0.37	0.0467 ± 0.0023		0.01451 ± 0.000603		0.0915 ± 0.0125		concordant	92.9 ± 3.9		88.9 ± 12.6	
2. 140523-2 no.29	5154	260	197992	3844	529994	0.40	0.0504 ± 0.0024		0.01506 ± 0.000637		0.1026 ± 0.0141		concordant	96.3 ± 4.1		99.2 ± 14.2	
2. 140523-2 no.30	5984	241	318992	3750	516994	0.67	0.0474 ± 0.0023		0.01522 ± 0.000646		0.0975 ± 0.0138		concordant	97.4 ± 4.2		94.5 ± 13.9	
													Average ± 2SD	95.1 ± 6.4		95.9 ± 23.2	
													Weighted average	94.9 ± 0.8		94.9 ± 2.7	

Th/U is ratio of thorium and uranium concentrations. Disc.* is degree of discordance. Positive and negative values mean left- and right-side significant offsets from concordia line, respectively.

Table 2c. U-Pb ages for adopted zircon grains obtained from Sp. 140523-2.

sample name	Total count 206Pb	207Pb	232Th	238U	238U	Th/U	Isotopic ratios 207Pb/206Pb	Error 2σ	206Pb/238U	Error 2σ	207Pb/235U	Error 2σ	Disc.*	Age (Ma) 206Pb/238U	Error 2σ	207Pb/206Pb	Error 2σ
2. 140523-2 no.1	3775	219	93593	2945	405994	0.25	0.0579	± 0.0017	0.01446	± 0.000644	0.1115	± 0.0160	concordant	92.5	± 4.1	107.4	± 16.1
2. 140523-2 no.2	5285	306	284993	3815	525994	0.59	0.0579	± 0.0017	0.01562	± 0.000640	0.1203	± 0.0148	concordant	99.9	± 4.1	115.4	± 14.9
2. 140523-2 no.3	2315	104	80693	1777	244994	0.36	0.0448	± 0.0013	0.01469	± 0.000757	0.0877	± 0.0179	concordant	94.0	± 4.9	85.3	± 18.0
2. 140523-2 no.4	6875	332	198993	5207	717994	0.30	0.0483	± 0.0014	0.01483	± 0.000677	0.0996	± 0.0112	concordant	95.3	± 3.7	92.7	± 11.3
2. 140523-2 no.5	2675	159	90393	2074	285994	0.35	0.0593	± 0.0018	0.01454	± 0.000716	0.1149	± 0.0192	concordant	93.1	± 4.6	110.4	± 19.4
2. 140523-2 no.6	804	44	33993	674	192894	0.40	0.0535	± 0.0015	0.01480	± 0.001095	0.0975	± 0.0306	concordant	94.7	± 7.1	94.5	± 30.6
2. 140523-2 no.7	2985	137	116993	2473	340094	0.38	0.0458	± 0.0014	0.01361	± 0.000716	0.0830	± 0.0148	concordant	87.1	± 4.2	81.0	± 14.9
2. 140523-2 no.8	4875	251	149993	3757	517994	0.32	0.0514	± 0.0015	0.01463	± 0.000648	0.1002	± 0.0134	concordant	93.6	± 3.9	97.0	± 13.5
2. 140523-2 no.9	5755	284	179993	4366	601994	0.33	0.0493	± 0.0015	0.01486	± 0.000611	0.0976	± 0.0123	concordant	95.1	± 3.8	94.5	± 12.4
2. 140523-2 no.10	5875	326	160993	4352	599994	0.29	0.0554	± 0.0016	0.01522	± 0.000597	0.1124	± 0.0134	concordant	97.4	± 3.9	108.1	± 13.5
2. 140523-2 no.11	7845	381	207993	6179	851994	0.27	0.0485	± 0.0014	0.01432	± 0.000609	0.0925	± 0.0123	concordant	91.6	± 3.8	89.8	± 10.3
2. 140523-2 no.12	4065	182	126993	3032	417994	0.33	0.0447	± 0.0013	0.01512	± 0.000597	0.0900	± 0.0134	concordant	96.7	± 3.5	87.5	± 14.1
2. 140523-2 no.13	2015	86	143993	1501	206994	0.76	0.0426	± 0.0013	0.01513	± 0.000660	0.0857	± 0.0102	concordant	96.8	± 4.3	83.5	± 19.3
2. 140523-2 no.14	4525	223	156993	3365	463994	0.34	0.0492	± 0.0015	0.01516	± 0.000640	0.0994	± 0.0140	concordant	97.0	± 4.2	96.2	± 14.2
2. 140523-2 no.15	6515	292	223993	5164	711994	0.45	0.0448	± 0.0013	0.01423	± 0.000557	0.0848	± 0.0141	concordant	91.1	± 3.5	82.7	± 10.6
2. 140523-2 no.16	1424	81	58692	1015	139994	0.45	0.0567	± 0.0028	0.01575	± 0.000977	0.1208	± 0.0192	concordant	100.7	± 5.3	115.8	± 28.6
2. 140523-2 no.17	1254	54	52892	921	126994	0.52	0.0429	± 0.0021	0.01529	± 0.000994	0.0686	± 0.0285	concordant	97.8	± 6.3	88.2	± 25.3
2. 140523-2 no.18	5214	279	249992	3757	517994	0.52	0.0535	± 0.0026	0.01558	± 0.000658	0.1127	± 0.0141	concordant	99.7	± 3.6	108.4	± 15.2
2. 140523-2 no.19	15874	843	773992	11804	1598994	0.45	0.0531	± 0.0026	0.01536	± 0.000646	0.1103	± 0.0105	concordant	98.3	± 6.4	106.2	± 25.3
2. 140523-2 no.20	1034	65	47192	827	113994	0.62	0.0628	± 0.0030	0.01404	± 0.000984	0.1189	± 0.0285	concordant	89.9	± 4.2	114.1	± 31.2
2. 140523-2 no.21	5214	240	368992	3887	535994	0.55	0.0460	± 0.0022	0.01506	± 0.000806	0.0837	± 0.0150	concordant	96.4	± 3.5	90.9	± 13.4
2. 140523-2 no.22	1624	96	98392	1255	172994	0.41	0.0590	± 0.0029	0.01453	± 0.000646	0.1159	± 0.0095	concordant	93.0	± 6.3	111.3	± 9.6
2. 140523-2 no.23	5794	262	235992	4460	614994	0.41	0.0500	± 0.0024	0.01459	± 0.000859	0.0994	± 0.0312	concordant	94.1	± 4.1	96.3	± 13.4
2. 140523-2 no.24	2174	109	116992	1661	228994	0.47	0.0452	± 0.0024	0.01438	± 0.000787	0.0891	± 0.0133	concordant	93.3	± 5.5	88.7	± 25.3
2. 140523-2 no.25	5794	143	36392	2205	303994	0.13	0.0492	± 0.0023	0.01451	± 0.000724	0.0983	± 0.0252	concordant	93.3	± 5.1	95.2	± 25.4
2. 140523-2 no.26	2904	186	187992	3104	427994	0.47	0.0467	± 0.0023	0.01438	± 0.000647	0.0909	± 0.0203	concordant	94.6	± 3.9	88.3	± 12.3
2. 140523-2 no.27	3974	263	207992	4359	600994	0.37	0.0486	± 0.0023	0.01451	± 0.000603	0.0915	± 0.0177	concordant	92.0	± 4.7	88.9	± 20.4
2. 140523-2 no.28	5834	260	197992	3844	528994	0.40	0.0504	± 0.0024	0.01506	± 0.000637	0.1026	± 0.0144	concordant	92.9	± 4.2	95.2	± 17.8
2. 140523-2 no.29	5154	253	207992	3104	528994	0.40	0.0504	± 0.0023	0.01506	± 0.000603	0.0915	± 0.0125	concordant	96.3	± 3.9	99.2	± 12.6
2. 140523-2 no.30	5084	241	318992	3750	516994	0.67	0.0474	± 0.0023	0.01522	± 0.000646	0.0875	± 0.0138	concordant	97.4	± 4.2	94.5	± 13.9
										Average ± 2SD				94.9	± 6.2	96.8	± 21.3
										Weighted average				94.8	± 0.8	95.3	± 2.7

Th/U is ratio of thorium and uranium concentrations. Disc.* is degree of discordance. Positive and negative values mean left- and right-side significant offsets from concordia line, respectively.

43

Table 3a. Fission-track grain ages of zircons obtained from Sp. 140523-3.

No.	Ns	Nu	S x10^{-5} (cm^2)	ρ_s x10^{-7} (cm^{-2})	ρ_u x10^{-9} (cm^{-2})	Ns / Nu x10^3	T (Ma)	σ_T (Ma)	U (ppm)
37	109	158	1.20	0.91	1.32	689.87	115.51	16.35	121
39	110	190	2.50	0.44	0.76	578.95	97.07	13.34	70
1	134	237	1.20	1.12	1.98	565.40	94.82	12.07	182
11	192	378	3.00	0.64	1.26	507.94	85.25	9.49	116
36	69	138	1.60	0.43	0.86	500.00	83.92	13.60	79
3	98	201	0.60	1.63	3.35	487.56	81.85	11.49	308
34	135	279	1.60	0.84	1.74	483.87	81.23	10.12	160
38	180	383	2.40	0.75	1.60	469.97	78.91	8.89	147
33	85	185	1.60	0.53	1.16	459.46	77.16	11.37	106
22	74	162	0.80	0.93	2.03	456.79	76.71	11.94	186
31	206	481	1.50	1.37	3.21	428.27	71.95	7.70	295
2	56	134	0.60	0.93	2.23	417.91	70.22	12.13	205
27	92	224	0.50	1.84	4.48	410.71	69.02	9.73	412
40	150	376	1.80	0.83	2.09	398.94	67.05	7.89	192
25	189	508	1.60	1.18	3.18	372.05	62.55	6.79	292
4	49	141	0.40	1.23	3.53	347.52	58.44	10.46	324
35	84	247	0.80	1.05	3.09	340.08	57.20	8.19	284
32	46	151	1.00	0.46	1.51	304.64	51.26	9.30	139
19	79	265	0.90	0.88	2.94	298.11	50.17	7.26	271

Ns is number of spontaneous tracks, Nu is number of ^{238}U counts, S is analyzed area of crystal, ρ_s is density of spontaneous tracks, ρ_u is density of ^{238}U counts, σ_T is error for each grain age (1σ), U is uranium density. Uranium concentration for standardization using 91500 standard zircon is 0.103x10^6 cm^2. Epsilon (ε) corresponding to conventional zeta (ζ) value for Fish Canyon Tuff zircon is 32.9±2.2.

The FT dating results for 26 zircon grains from Sp. 140523-2 are summarized in Table 2a. As shown in Figure 14a, the grain ages are well grouped and their average age is 71.1±4.8 Ma (error: 1σ). Figure 14b indicates that the U-Pb ages for the sample are significantly older than its FT ages.

The U-Pb dating results for total 30 zircon grains from Sp. 140523-2 are summarized in Table 2b. As shown in the concordia plot of the U-Pb zircon ages (Figure 14c), a large proportion of the dated grains fall close to the concordia line. We excluded one grain that had discordant age. The final results are presented in Table 2c and Figure 14d. The average age of the selected grains is 94.8±0.8 Ma (error: 2σ).

Table 3b. U-Pb ages for all analyzed zircon grains obtained from Sp. 140523-3.

sample name	Total count 206Pb	207Pb	232Th	238U	Isotopic ratios 238U	Th/U	207Pb/206Pb	Error 2σ	206Pb/238U	Error 2σ	207Pb/235U	Error 2σ	Disc.*	Age (Ma) 206Pb/238U	Error 2σ	207Pb/206Pb	Error 2σ
3. 140523-3 no.1	3385	160	275988	2821	388989	0.80	0.0473	0.0032	0.01434	0.000785	0.0928	0.0158	concordant	91.8	5.1	90.1	16.0
3. 140523-3 no.2	1075	51	25288	928	127989	0.22	0.0475	0.0032	0.01384	0.001032	0.0900	0.0263	concordant	88.6	6.7	87.5	26.4
3. 140523-3 no.3	2675	102	78348	2176	299989	0.29	0.0381	0.0025	0.01470	0.000846	0.0768	0.0161	concordant	94.0	5.5	75.1	16.2
3. 140523-3 no.4	3805	176	132988	3024	416989	0.36	0.0463	0.0031	0.01504	0.000805	0.0953	0.0156	concordant	96.2	5.2	92.4	15.7
3. 140523-3 no.5	2815	136	118988	2154	296989	0.45	0.0463	0.0032	0.01562	0.000889	0.1034	0.0191	concordant	99.9	5.7	99.9	19.2
3. 140523-3 no.6	3455	184	394988	2836	390989	1.13	0.0533	0.0036	0.01456	0.000758	0.1062	0.0171	concordant	93.2	5.1	102.5	17.2
3. 140523-3 no.7	5185	233	182988	4156	572989	0.36	0.0449	0.0030	0.01491	0.000851	0.0918	0.0132	concordant	95.4	4.9	89.2	13.4
3. 140523-3 no.8	2555	108	149988	2089	287989	0.58	0.0423	0.0028	0.01462	0.000704	0.0847	0.0173	concordant	93.6	5.5	82.5	17.4
3. 140523-3 no.9	7005	336	403988	5609	800989	0.57	0.0490	0.0032	0.01441	0.000914	0.0947	0.0117	concordant	92.2	4.5	91.8	11.8
3. 140523-3 no.10	1615	75	115988	1385	190989	0.68	0.0465	0.0031	0.01394	0.000944	0.0887	0.0215	concordant	89.2	5.9	86.3	21.6
3. 140523-3 no.11	1675	93	86988	1378	189989	0.51	0.0555	0.0037	0.01453	0.000971	0.1105	0.0244	-41	93.0	6.1	106.4	24.4
3. 140523-3 no.12	2025	183	63788	1537	211989	0.34	0.0604	0.0060	0.01574	0.000677	0.1298	0.0321	concordant	100.7	6.3	190.7	32.1
3. 140523-3 no.13	14375	717	440988	11604	599989	0.86	0.0499	0.0033	0.01513	0.000779	0.0948	0.0094	-5	96.8	5.0	97.8	9.5
3. 140523-3 no.14	4775	299	399988	3871	519989	0.31	0.0626	0.0042	0.01472	0.000678	0.1123	0.0170	concordant	94.2	4.4	123.9	17.1
3. 140523-3 no.15	12775	640	587988	10371	1429989	0.46	0.0601	0.0033	0.01486	0.000740	0.1010	0.0098	concordant	95.1	4.4	97.7	9.9
3. 140523-3 no.16	1193	56	43685	957	131985	0.37	0.0465	0.0028	0.01487	0.001063	0.0933	0.0265	concordant	95.2	6.8	90.6	26.5
3. 140523-3 no.17	5263	296	214985	4235	583985	0.41	0.0561	0.0034	0.01483	0.000744	0.1123	0.0153	concordant	94.9	4.8	108.0	15.5
3. 140523-3 no.18	3693	198	165985	3140	432985	0.43	0.0536	0.0027	0.01403	0.000465	0.1012	0.0162	concordant	89.8	4.8	97.9	16.3
3. 140523-3 no.19	1603	72	55985	1305	179985	0.35	0.0446	0.0027	0.01465	0.000680	0.0881	0.0222	concordant	93.8	6.1	85.8	22.3
3. 140523-3 no.20	7273	386	393985	5795	798985	0.55	0.0530	0.0032	0.01497	0.000715	0.1071	0.0132	concordant	95.8	4.6	103.3	13.3
3. 140523-3 no.21	6473	312	960985	5505	758985	1.42	0.0481	0.0029	0.01403	0.000680	0.0911	0.0121	concordant	89.8	4.4	91.9	12.2
3. 140523-3 no.22	3193	150	124985	2538	349985	0.40	0.0468	0.0026	0.01501	0.000820	0.0948	0.0171	concordant	96.0	5.3	78.5	17.2
3. 140523-3 no.23	6433	273	433985	5454	751985	0.65	0.0424	0.0026	0.01407	0.000682	0.0804	0.0113	concordant	90.1	4.4	104.3	11.4
3. 140523-3 no.24	7193	376	279985	5584	769985	0.41	0.0522	0.0031	0.01537	0.000734	0.1062	0.0135	concordant	98.3	4.7	74.3	13.6
3. 140523-3 no.25	2113	83	104985	1748	240985	0.49	0.0390	0.0024	0.01443	0.000869	0.0760	0.0178	concordant	92.4	5.6	94.5	17.9
3. 140523-3 no.26	5213	254	490985	4185	530985	0.89	0.0466	0.0029	0.01486	0.000743	0.0975	0.0141	concordant	95.1	4.8	88.3	14.2
3. 140523-3 no.27	4893	218	239985	3851	530985	0.51	0.0444	0.0027	0.01516	0.000765	0.0909	0.0140	concordant	97.0	4.9	100.6	14.1
3. 140523-3 no.28	2623	138	66005	2125	292985	0.25	0.0504	0.0032	0.01473	0.000841	0.1041	0.0195	concordant	94.3	5.1	101.8	19.6
3. 140523-3 no.29	3193	174	163985	2647	364985	0.50	0.0543	0.0033	0.01439	0.000787	0.1055	0.0179	concordant	92.1	5.1	104.5	18.0
3. 140523-3 no.30	1323	70	67685	1023	140985	0.54	0.0525	0.0032	0.01544	0.001067	0.1094	0.0280	concordant	98.6	6.9	105.4	28.1
Average ± 2SD														94.2 ± 6.1		97.3 ± 37.9	
Weighted average														94.1 ± 0.9		94.8 ± 2.8	

Th/U is ratio of thorium and uranium concentrations. Disc.* is degree of discordance. Positive and negative values mean left- and right-side significant offsets from concordia line, respectively.

45

Table 3c. U-Pb ages for adopted zircon grains obtained from Sp. 140523-3.

sample name	Total count					Th/U	Isotopic ratios						Disc.*	Age (Ma)			
	206Pb	207Pb	232Th	235U	238U		207Pb/206Pb	Error 2σ	206Pb/238U	Error 2σ	207Pb/235U	Error 2σ		206Pb/238U	Error 2σ	207Pb/235U	Error 2σ
3. 140523-3 no.1	3385	160	275988	2821	386989	0.80	0.04728	±0.00316	0.01434	±0.00079	0.09264	±0.01584	concordant	91.8	±5.1	90.1	±16.0
3. 140523-3 no.2	1075	51	25288	928	127989	0.22	0.04748	±0.00317	0.01384	±0.00103	0.09000	±0.02630	concordant	88.6	±6.7	87.5	±26.4
3. 140523-3 no.3	2675	102	78388	2176	299989	0.29	0.03815	±0.00255	0.01470	±0.00085	0.07676	±0.01605	concordant	94.0	±5.5	75.1	±16.2
3. 140523-3 no.4	3805	176	132988	3024	416989	0.36	0.04627	±0.00309	0.01504	±0.00061	0.09526	±0.01558	concordant	96.2	±5.2	92.4	±15.7
3. 140523-3 no.5	2815	136	118988	2154	296989	0.45	0.04833	±0.00323	0.01562	±0.00069	0.10336	±0.01905	concordant	99.9	±5.7	99.9	±19.2
3. 140523-3 no.6	3455	184	394988	2836	390989	1.13	0.05327	±0.00356	0.01456	±0.00079	0.10621	±0.01708	concordant	93.2	±5.1	102.5	±17.2
3. 140523-3 no.7	5185	233	182988	4156	572989	0.36	0.04494	±0.00300	0.01491	±0.00076	0.09177	±0.01325	concordant	95.4	±4.9	89.2	±13.4
3. 140523-3 no.8	2555	108	149988	2089	287989	0.58	0.04229	±0.00282	0.01462	±0.00085	0.08465	±0.01728	concordant	93.6	±5.5	82.5	±17.4
3. 140523-3 no.9	7005	336	403988	5809	800989	0.57	0.04797	±0.00320	0.01441	±0.00070	0.09466	±0.01171	concordant	92.2	±4.5	91.8	±11.8
3. 140523-3 no.10	1615	75	115988	1385	190989	0.68	0.04646	±0.00310	0.01394	±0.00091	0.08866	±0.02152	concordant	89.2	±5.9	86.3	±21.6
3. 140523-3 no.11	1675	93	86988	1378	189989	0.51	0.05554	±0.00371	0.01453	±0.00094	0.11051	±0.02436	concordant	93.0	±6.1	106.4	±24.4
3. 140523-3 no.13	14375	717	440988	11604	1599989	0.31	0.04988	±0.00333	0.01481	±0.00068	0.10112	±0.00940	concordant	94.7	±4.4	97.8	±9.5
3. 140523-3 no.14	12775	640	587988	10371	1429989	0.46	0.05010	±0.00335	0.01472	±0.00068	0.10099	±0.00977	concordant	94.2	±4.4	97.7	±9.9
3. 140523-3 no.16	1193	56	43685	957	131986	0.37	0.04651	±0.00280	0.01487	±0.00106	0.09331	±0.02646	concordant	95.2	±6.8	90.6	±26.5
3. 140523-3 no.17	5263	296	214985	4235	583986	0.41	0.05614	±0.00338	0.01483	±0.00074	0.11228	±0.01534	concordant	94.9	±4.8	108.0	±15.5
3. 140523-3 no.18	3693	198	165985	3140	432985	0.43	0.05548	±0.00322	0.01403	±0.00074	0.10121	±0.01623	concordant	89.8	±4.8	97.9	±16.3
3. 140523-3 no.19	1603	72	55985	1305	179985	0.35	0.04460	±0.00269	0.01465	±0.00095	0.08815	±0.02216	concordant	93.8	±6.1	85.8	±22.3
3. 140523-3 no.20	7273	386	393985	5795	798985	0.55	0.05300	±0.00320	0.01497	±0.00071	0.10706	±0.01322	concordant	95.8	±4.6	103.3	±13.3
3. 140523-3 no.21	6473	312	960985	5505	758985	1.42	0.04812	±0.00290	0.01403	±0.00068	0.09107	±0.01213	concordant	89.8	±4.4	88.5	±12.2
3. 140523-3 no.22	3193	150	124985	2538	349985	0.40	0.04662	±0.00282	0.01501	±0.00082	0.09478	±0.01709	concordant	96.0	±5.3	91.9	±17.2
3. 140523-3 no.23	6433	273	433985	5454	751985	0.65	0.04226	±0.00255	0.01407	±0.00068	0.08941	±0.01126	concordant	90.1	±4.4	78.5	±11.4
3. 140523-3 no.24	7193	376	279985	5584	769985	0.41	0.05220	±0.00315	0.01537	±0.00073	0.10821	±0.01349	concordant	98.3	±4.7	104.3	±13.6
3. 140523-3 no.25	2113	83	104985	1748	240985	0.49	0.03904	±0.00235	0.01443	±0.00087	0.07596	±0.01781	concordant	92.3	±5.6	74.3	±17.9
3. 140523-3 no.26	5213	254	460985	4185	576985	0.89	0.04863	±0.00293	0.01486	±0.00074	0.09749	±0.01410	concordant	95.1	±4.8	94.5	±14.2
3. 140523-3 no.27	4893	218	239985	3851	530985	0.51	0.04445	±0.00268	0.01516	±0.00077	0.09089	±0.01397	concordant	97.0	±4.9	88.3	±14.1
3. 140523-3 no.28	2623	138	66085	2125	292985	0.25	0.05241	±0.00316	0.01473	±0.00084	0.10413	±0.01953	concordant	94.3	±5.4	100.6	±19.6
3. 140523-3 no.29	3193	174	163985	2647	364985	0.50	0.05433	±0.00328	0.01439	±0.00079	0.10548	±0.01788	concordant	92.1	±5.1	101.8	±18.0
3. 140523-3 no.30	1323	70	67685	1023	140985	0.54	0.05252	±0.00317	0.01544	±0.00107	0.10938	±0.02803	concordant	98.8	±6.9	105.4	±28.1
												Average ± 2SD		93.9	±5.7	93.3	±18.6
												Weighted average		93.8	±1.0	93.3	±2.9

Th/U is ratio of thorium and uranium concentrations. Disc.* is degree of discordance. Positive and negative values mean left- and right-side significant offsets from concordia line, respectively.

Table 4a. Fission-track grain ages of zircons obtained from Sp. 140523-1.

No.	Ns	Nu	S x10^{-5} (cm^2)	ρ_s x10^{-7} (cm^{-2})	ρ_u x10^{-9} (cm^{-2})	Ns / Nu x10^3	T (Ma)	σ_T (Ma)	U (ppm)
36	127	214	1.00	1.27	2.14	593.46	99.49	13.00	197
12	58	103	0.80	0.73	1.29	563.11	94.44	16.75	118
6	107	207	1.20	0.89	1.73	516.91	86.74	11.86	159
24	75	146	0.60	1.25	2.43	513.70	86.21	13.55	224
30	51	110	1.40	0.36	0.79	463.64	77.86	14.19	72
4	172	386	1.00	1.72	3.86	445.60	74.84	8.51	355
33	97	218	1.00	0.97	2.18	444.95	74.74	10.42	200
10	38	96	0.20	1.90	4.80	395.83	66.53	13.51	441
35	94	250	0.60	1.57	4.17	376.00	63.21	8.75	383
34	173	486	1.20	1.44	4.05	355.97	59.86	6.66	372
2	73	211	0.60	1.22	3.52	345.97	58.19	8.82	323
18	188	579	1.20	1.57	4.83	324.70	54.62	5.88	443
20	84	272	0.60	1.40	4.53	308.82	51.96	7.37	417
21	116	433	1.20	0.97	3.61	267.90	45.10	5.61	332
9	47	201	0.30	1.57	6.70	233.83	39.38	6.91	616

Ns is number of spontaneous tracks, Nu is number of ^{238}U counts, S is analyzed area of crystal, ρ_s is density of spontaneous tracks, ρ_u is density of ^{238}U counts, σ_T is error for each grain age (1σ), U is uranium density. Uranium concentration for standardization using 91500 standard zircon is 0.103x10^6 cm^2. Epsilon (ε) corresponding to conventional zeta (ζ) value for Fish Canyon Tuff zircon is 32.9±2.2.

The FT dating results for 19 zircon grains from Sp. 140523-3 are summarized in Table 3a. As shown in Figure 14a, the grain ages are well grouped and their average age is 74.2±5.4 Ma (error: 1σ). Figure 14b indicates that the U-Pb ages for the sample are significantly older than its FT ages.

The U-Pb dating results for total 30 zircon grains from Sp. 140523-3 are summarized in Table 3b. Based on the concordia plot for U-Pb zircon ages in Figure 14c, we excluded two grains with discordant ages. The final results are presented in Table 3c and Figure 14d. The average age of the selected grains is 93.8±1.0 Ma (error: 2σ).

The FT dating results for 15 zircon grains from Sp. 140523-1 are summarized in Table 4a. As shown in Figure 14a, the grain ages are fairly scattered, with the average age being 64.5±4.8 Ma (error: 1σ). Figure 14b indicates that the U-Pb ages for the sample are significantly older than its FT ages.

Table 4b. U-Pb ages for all analyzed zircon grains obtained from Sp. 140523-1.

sample name	Total count					Th/U	Isotopic ratios						Disc.*	Age (Ma)			
	206Pb	207Pb	232Th	235U	238U		207Pb/206Pb	Error 2σ	206Pb/238U	Error 2σ	207Pb/235U	Error 2σ		206Pb/238U	Error 2σ	207Pb/235U	Error 2σ
140523-1 no.1	7089	329	271943	4895	674949	0.45	0.0463	0.0031	0.01587	0.000757	0.0979	0.0120	concordant	101.5 ± 4.9		94.8 ± 12.1	
140523-1 no.2	4169	162	297943	3082	424949	0.76	0.0387	0.0026	0.01482	0.000767	0.0764	0.0128	3	94.9 ± 4.9		74.8 ± 12.9	
140523-1 no.3	5649	267	296943	4213	580949	0.57	0.0472	0.0031	0.01469	0.000723	0.0923	0.0124	concordant	94.0 ± 4.7		89.6 ± 12.5	
140523-1 no.4	5099	204	179943	3858	531949	0.38	0.0399	0.0027	0.01448	0.000724	0.0769	0.0116	1	92.7 ± 4.7		75.3 ± 11.7	
140523-1 no.5	12429	571	688943	8775	1206949	0.63	0.0459	0.0031	0.01552	0.000699	0.0948	0.0093	concordant	99.3 ± 4.5		92.0 ± 9.4	
140523-1 no.6	4809	236	289943	3532	496949	0.66	0.0490	0.0033	0.01492	0.000753	0.0973	0.0138	concordant	95.5 ± 4.9		94.3 ± 14.0	
140523-1 no.7	3619	145	240943	2531	348949	0.77	0.0399	0.0027	0.01567	0.000833	0.0833	0.0148	concordant	100.2 ± 5.4		81.2 ± 14.9	
140523-1 no.8	6639	446	256943	6027	830949	0.34	0.0516	0.0034	0.01571	0.000732	0.1078	0.0117	concordant	100.5 ± 4.7		104.0 ± 11.8	
140523-1 no.9	6779	372	338943	4736	652949	0.58	0.0548	0.0036	0.01569	0.000753	0.1144	0.0134	concordant	100.3 ± 4.9		110.0 ± 13.5	
140523-1 no.10	3819	284	177943	2908	400949	0.49	0.0742	0.0049	0.01439	0.000757	0.1422	0.0189	-21	92.1 ± 4.9		135.0 ± 19.0	
140523-1 no.11	7089	366	204943	5142	708949	0.32	0.0516	0.0034	0.01511	0.000721	0.1037	0.0122	concordant	96.7 ± 4.6		100.2 ± 12.3	
140523-1 no.12	2179	100	64143	1573	216949	0.33	0.0457	0.0030	0.01518	0.000606	0.0922	0.0195	concordant	97.1 ± 5.8		89.6 ± 19.6	
140523-1 no.13	9889	498	441943	7325	1009949	0.49	0.0503	0.0033	0.01478	0.000680	0.0991	0.0102	concordant	94.7 ± 4.4		95.9 ± 10.4	
140523-1 no.14	5079	227	280943	3379	465949	0.67	0.0446	0.0030	0.01647	0.000824	0.0978	0.0142	concordant	105.3 ± 5.3		94.7 ± 14.3	
140523-1 no.15	4949	249	97643	3155	434949	0.25	0.0502	0.0033	0.01719	0.000864	0.1149	0.0160	concordant	109.9 ± 5.6		110.6 ± 16.2	
140523-1 no.16	5995	263	265990	4228	582992	0.50	0.0439	0.0028	0.01581	0.000794	0.0937	0.0127	concordant	101.1 ± 5.1		91.0 ± 12.8	
140523-1 no.17	1125	47	74790	819	112992	0.73	0.0419	0.0027	0.01531	0.001129	0.0866	0.0262	concordant	97.9 ± 7.3		84.3 ± 26.3	
140523-1 no.18	4775	245	115990	3561	490992	0.26	0.0513	0.0033	0.01495	0.000776	0.1037	0.0145	concordant	95.7 ± 5.0		100.1 ± 14.6	
140523-1 no.19	7345	370	301990	5505	758992	0.44	0.0504	0.0033	0.01488	0.000728	0.1013	0.0118	concordant	95.2 ± 4.7		97.9 ± 11.9	
140523-1 no.20	5055	252	312990	3793	522992	0.66	0.0499	0.0032	0.01486	0.000764	0.1001	0.0138	concordant	95.1 ± 4.9		96.9 ± 13.9	
140523-1 no.21	1435	64	27990	1081	148992	0.21	0.0447	0.0029	0.01481	0.001011	0.0893	0.0233	concordant	94.8 ± 6.5		86.9 ± 23.4	
140523-1 no.22	5605	264	181990	4497	619992	0.32	0.0471	0.0030	0.01390	0.000704	0.0885	0.0119	concordant	89.0 ± 4.5		86.1 ± 12.0	
140523-1 no.23	4835	228	214990	3590	494992	0.48	0.0472	0.0030	0.01502	0.000778	0.0967	0.0138	concordant	96.1 ± 5.0		92.8 ± 13.9	
140523-1 no.24	2125	107	74990	1625	223992	0.37	0.0504	0.0033	0.01459	0.000893	0.0993	0.0203	concordant	93.4 ± 5.8		96.1 ± 20.4	
140523-1 no.25	3935	170	136990	2749	373992	0.40	0.0432	0.0028	0.01596	0.000866	0.0932	0.0153	concordant	102.1 ± 5.5		90.5 ± 15.5	
140523-1 no.26	7345	355	683990	5555	766992	0.99	0.0453	0.0028	0.01474	0.000722	0.0963	0.0114	concordant	94.3 ± 4.7		93.3 ± 11.5	
140523-1 no.27	2445	106	130990	1849	254992	0.57	0.0454	0.0028	0.01474	0.000872	0.0864	0.0177	concordant	94.3 ± 5.6		84.1 ± 17.8	
140523-1 no.28	14075	711	296990	5650	778992	0.42	0.0505	0.0033	0.02776	0.001284	0.1896	0.0174	concordant	176.6 ± 8.3		176.2 ± 17.6	
140523-1 no.29	15175	726	608990	11604	1599992	0.42	0.0478	0.0031	0.01458	0.000670	0.0942	0.0084	concordant	93.3 ± 4.3		91.4 ± 8.5	
140523-1 no.30	4965	259	241990	3663	504992	0.53	0.0522	0.0034	0.01512	0.000780	0.1065	0.0146	concordant	96.7 ± 5.0		102.8 ± 14.7	
													Average ± 2SD	99.7 ± 30.3		97.1 ± 37.6	
													Weighted average	97.8 ± 0.9		95.3 ± 2.4	

Th/U is ratio of thorium and uranium concentrations. Disc.* is degree of discordance. Positive and negative values mean left- and right-side significant offsets from concordia line, respectively.

Table 4c. U-Pb ages for adopted zircon grains obtained from Sp. 140523-1.

sample name	Total count					Th/U	Isotopic ratios						Disc.*	Age (Ma)			
	206Pb	207Pb	232Th	238U	238U		207Pb/206Pb	Error 2σ	206Pb/238U	Error 2σ	207Pb/235U	Error 2σ		206Pb/238U	Error 2σ	207Pb/235U	Error 2σ
1. 140523-1 no.1	7089	329	271943	4895	674949	0.45	0.04634 ± 0.00308	0.01587 ± 0.00076	0.09789 ± 0.01203				concordant	101.5 ± 4.9		94.8 ± 12.1	
1. 140523-1 no.3	5649	267	296943	4213	580949	0.57	0.04717 ± 0.00314	0.01469 ± 0.00072	0.09227 ± 0.01240				concordant	94.0 ± 4.7		89.6 ± 12.5	
1. 140523-1 no.5	12429	571	688943	8775	1209949	0.63	0.04590 ± 0.00305	0.01552 ± 0.00070	0.09484 ± 0.00928				concordant	99.3 ± 4.5		92.0 ± 9.4	
1. 140523-1 no.6	4809	238	289943	3532	486949	0.66	0.04897 ± 0.00326	0.01492 ± 0.00075	0.09727 ± 0.01383				concordant	95.5 ± 4.9		94.3 ± 14.0	
1. 140523-1 no.7	3619	145	240943	2531	348949	0.77	0.03992 ± 0.00266	0.01567 ± 0.00063	0.08329 ± 0.01475				concordant	100.2 ± 5.4		81.2 ± 14.9	
1. 140523-1 no.8	8639	446	256943	6027	830949	0.34	0.05157 ± 0.00343	0.01571 ± 0.00073	0.10784 ± 0.01169				concordant	100.5 ± 4.7		104.0 ± 11.8	
1. 140523-1 no.9	6779	372	338943	4736	662949	0.58	0.05480 ± 0.00365	0.01569 ± 0.00075	0.11444 ± 0.01340				concordant	100.3 ± 4.9		110.0 ± 13.5	
1. 140523-1 no.11	7089	366	204943	5142	708949	0.32	0.05156 ± 0.00343	0.01511 ± 0.00072	0.10370 ± 0.01219				concordant	96.7 ± 4.6		100.2 ± 12.3	
1. 140523-1 no.12	2179	100	64143	1573	216949	0.33	0.04566 ± 0.00304	0.01518 ± 0.00075	0.09225 ± 0.01954				concordant	97.1 ± 5.8		89.6 ± 19.6	
1. 140523-1 no.13	9889	498	441943	7325	1009949	0.49	0.05031 ± 0.00335	0.01479 ± 0.00068	0.09908 ± 0.01025				concordant	94.7 ± 4.4		96.9 ± 10.4	
1. 140523-1 no.14	5079	227	280943	3379	465949	0.67	0.04459 ± 0.00297	0.01647 ± 0.00082	0.09777 ± 0.01415				concordant	105.3 ± 5.3		94.7 ± 14.3	
1. 140523-1 no.16	5995	263	265990	4228	582992	0.50	0.04389 ± 0.00283	0.01581 ± 0.00079	0.09371 ± 0.01267				concordant	101.1 ± 5.1		91.0 ± 12.8	
1. 140523-1 no.17	1125	47	74790	819	112992	0.73	0.04187 ± 0.00270	0.01531 ± 0.00113	0.08656 ± 0.02625				concordant	97.9 ± 7.3		84.3 ± 26.3	
1. 140523-1 no.18	4775	245	116990	3561	490992	0.26	0.05133 ± 0.00332	0.01495 ± 0.00078	0.10386 ± 0.01450				concordant	95.7 ± 5.0		100.1 ± 14.6	
1. 140523-1 no.19	7345	370	301990	5505	758992	0.44	0.05039 ± 0.00325	0.01488 ± 0.00073	0.10125 ± 0.01184				concordant	95.2 ± 4.7		97.9 ± 11.9	
1. 140523-1 no.20	5055	252	312990	3793	522992	0.66	0.04987 ± 0.00322	0.01486 ± 0.00076	0.10009 ± 0.01362				concordant	95.1 ± 4.9		96.9 ± 13.9	
1. 140523-1 no.21	1435	64	27990	1081	148992	0.21	0.04467 ± 0.00288	0.01481 ± 0.00101	0.08934 ± 0.02334				concordant	94.8 ± 6.5		86.9 ± 23.4	
1. 140523-1 no.22	5605	264	181990	4497	619992	0.32	0.04712 ± 0.00304	0.01380 ± 0.00070	0.08845 ± 0.01192				concordant	89.0 ± 4.5		86.1 ± 12.0	
1. 140523-1 no.23	4835	228	214990	3590	494992	0.48	0.04718 ± 0.00305	0.01502 ± 0.00078	0.09569 ± 0.01380				concordant	96.1 ± 5.0		92.8 ± 13.9	
1. 140523-1 no.24	2125	107	74990	1625	223992	0.37	0.05040 ± 0.00326	0.01459 ± 0.00089	0.09929 ± 0.02033				concordant	93.4 ± 5.8		96.1 ± 20.4	
1. 140523-1 no.25	3935	170	136990	2749	378992	0.40	0.04323 ± 0.00279	0.01596 ± 0.00086	0.09320 ± 0.01534				concordant	102.1 ± 5.5		90.5 ± 16.5	
1. 140523-1 no.26	7345	355	683990	5555	765992	0.99	0.04835 ± 0.00312	0.01474 ± 0.00072	0.09626 ± 0.01144				concordant	94.3 ± 4.7		93.3 ± 11.5	
1. 140523-1 no.27	2445	106	130990	1849	254992	0.57	0.04339 ± 0.00280	0.01474 ± 0.00087	0.08640 ± 0.01771				concordant	94.3 ± 5.6		84.1 ± 17.8	
1. 140523-1 no.29	15175	726	608990	11604	1599992	0.42	0.04785 ± 0.00309	0.01458 ± 0.00067	0.09423 ± 0.00842				concordant	93.3 ± 4.3		91.4 ± 8.5	
1. 140523-1 no.30	4965	259	241990	3663	504992	0.53	0.05219 ± 0.00337	0.01512 ± 0.00078	0.10654 ± 0.01456				concordant	96.7 ± 5.0		102.8 ± 14.7	
Average ± 2SD														97.0 ± 7.2		93.6 ± 13.4	
Weighted average														96.8 ± 1.0		94.2 ± 2.6	

Th/U is ratio of thorium and uranium concentrations. Disc.* is degree of discordance. Positive and negative values mean left- and right-side significant offsets from concordia line, respectively.

49

Table 5a. Fission-track grain ages of zircons obtained from Sp. 121218-2.

No.	Ns	Nu	S x10⁻⁵ (cm²)	ρ_s x10⁻⁷ (cm⁻²)	ρ_u x10⁻⁹ (cm⁻²)	Ns / Nu x10³	T (Ma)	σ_T (Ma)	U (ppm)
20	125	2469	1.20	1.04	2.06	50.63	123.24	14.77	167
14	135	3175	1.50	0.90	2.12	42.52	103.66	12.12	172
7	226	5400	3.00	0.75	1.80	41.85	102.04	10.49	146
25	110	2664	1.20	0.92	2.22	41.29	100.69	12.51	180
15	144	3531	1.20	1.20	2.94	40.78	99.46	11.42	239
13	258	6672	2.00	1.29	3.34	38.67	94.34	9.43	271
10	102	2720	1.60	0.64	1.70	37.50	91.51	11.62	138
8	438	11767	2.40	1.83	4.90	37.22	90.84	8.29	398
1	215	6103	1.60	1.34	3.81	35.23	86.00	8.93	310
4	266	7995	2.40	1.11	3.33	33.27	81.25	8.06	271
17	152	4570	0.80	1.90	5.71	33.26	81.23	9.17	464
9	163	4907	1.50	1.09	3.27	33.22	81.12	9.00	266
19	47	1419	0.40	1.18	3.55	33.12	80.89	13.52	288
16	201	6113	1.80	1.12	3.40	32.88	80.31	8.46	276
22	118	3599	1.00	1.18	3.60	32.79	80.08	9.71	292
11	108	3306	1.20	0.90	2.76	32.67	79.79	9.94	224
30	148	4546	1.00	1.48	4.55	32.56	79.52	9.04	369
24	200	6174	1.20	1.67	5.15	32.39	79.12	8.34	418
2	105	3372	0.90	1.17	3.75	31.14	76.08	9.56	304
27	100	3480	0.80	1.25	4.35	28.74	70.24	8.95	353
6	270	9441	1.80	1.50	5.25	28.60	69.90	6.91	426
26	127	4488	1.40	0.91	3.21	28.30	69.17	8.20	260
18	136	4817	0.80	1.70	6.02	28.23	69.02	8.02	489
23	117	4188	1.60	0.73	2.62	27.94	68.30	8.29	213
5	89	3254	0.60	1.48	5.42	27.35	66.87	8.85	441
29	251	9626	2.40	1.05	4.01	26.08	63.77	6.39	326
12	135	5628	1.20	1.13	4.69	23.99	58.68	6.83	381
21	57	2398	0.40	1.43	6.00	23.77	58.15	8.99	487
3	264	11134	3.20	0.83	3.48	23.71	58.01	5.75	283
28	131	5889	2.40	0.55	2.45	22.24	54.44	6.39	199

Ns is number of spontaneous tracks, Nu is number of [238]U counts, S is analyzed area of crystal, ρ_s is density of spontaneous tracks, ρ_u is density of [238]U counts, σ_T is error for each grain age (1σ), U is uranium density. Uranium concentration for standardization using 91500 standard zircon is 1.162×10^6 cm². Epsilon (ε) corresponding to conventional zeta (ζ) value for Fish Canyon Tuff zircon is 42.3±2.4.

The U-Pb dating results for total 30 zircon grains from Sp. 140523-1 are summarized in Table 4b. Based on the concordia plot for U-Pb zircon ages in Figure 14c, we excluded three grains with discordant ages. Two other somewhat older grains were omitted as inherited zircons. The final results are presented in Table 4c and Figure 14d. The average age of the selected grains is 96.8±1.0 Ma (error: 2σ).

Table 5b. U–Pb ages for all analyzed zircon grains obtained from Sp. 121218-2.

sample name	Total count				Isotopic ratios						Disc.*	Age (Ma)			
	206Pb	207Pb	238U	235U	207Pb/206Pb	Error 2σ	206Pb/238U	Error 2σ	207Pb/235U	Error 2σ		206Pb/238U	Error 2σ	207Pb/235U	Error 2σ
121218-2 no.1	20353	1367	1197648	6686	0.0672 ± 0.0025	0.01726 ± 0.000530	0.1586 ± 0.0126	-20.9	110.3 ± 3.4	149.5 ± 12.7
121218-2 no.2	20415	1582	1176472	8533	0.0775 ± 0.0028	0.01762 ± 0.000541	0.1869 ± 0.0144	-38.5	112.6 ± 3.5	174.0 ± 14.5
121218-2 no.3	16706	1613	1092486	7923	0.0966 ± 0.0035	0.01553 ± 0.000487	0.2052 ± 0.0157	-71.6	99.3 ± 3.1	189.5 ± 15.9
121218-2 no.4	18268	778	1046006	7586	0.0426 ± 0.0016	0.01774 ± 0.000551	0.1034 ± 0.0098	0.2	113.3 ± 3.6	99.9 ± 9.7
121218-2 no.5	26245	1983	1703022	12351	0.0796 ± 0.0028	0.01565 ± 0.000469	0.1618 ± 0.0117	-37.3	100.1 ± 3.0	152.3 ± 11.8
121218-2 no.6	25169	1403	1646875	11944	0.0557 ± 0.0020	0.01552 ± 0.000466	0.1184 ± 0.0092	-2.0	99.3 ± 3.0	113.6 ± 9.3
121218-2 no.7(2)	9137	386	565201	4099	0.0425 ± 0.0015	0.01642 ± 0.000566	0.0954 ± 0.0086	concordant	105.0 ± 3.6	92.5 ± 11.5
121218-2 no.8(2)	22605	1161	1539560	11166	0.0509 ± 0.0019	0.01504 ± 0.000496	0.1048 ± 0.0086	concordant	96.3 ± 2.9	101.2 ± 8.7
121218-2 no.9	15204	834	1027262	7451	0.0548 ± 0.0020	0.01503 ± 0.000477	0.1128 ± 0.0102	concordant	96.2 ± 3.1	108.5 ± 10.3
121218-2 no.10(2)	8012	425	535831	3872	0.0530 ± 0.0019	0.01524 ± 0.000539	0.1106 ± 0.0128	concordant	97.5 ± 3.5	106.5 ± 12.9
121218-2 no.11	16816	1766	8850032	6274	0.1051 ± 0.0038	0.01974 ± 0.000619	0.2940 ± 0.0216	-81.0	126.0 ± 4.0	253.9 ± 21.7
121218-2 no.12	22208	2388	14472716	10681	0.1075 ± 0.0039	0.01532 ± 0.000466	0.2253 ± 0.0159	-91.2	98.0 ± 3.0	206.4 ± 16.0
121218-2 no.13	18497	1544	1047532	7597	0.0835 ± 0.0030	0.01793 ± 0.000556	0.2048 ± 0.0159	-48.0	114.6 ± 3.6	189.2 ± 16.0
121218-2 no.14	10338	597	664554	4820	0.0577 ± 0.0021	0.01580 ± 0.000532	0.1248 ± 0.0128	-2.1	101.1 ± 3.4	119.4 ± 12.9
121218-2 no.15	15707	1226	924004	6702	0.0781 ± 0.0028	0.01726 ± 0.000546	0.1844 ± 0.0152	-38.7	110.3 ± 3.5	171.8 ± 15.3
121218-2 no.16	15558	770	1066452	7735	0.0492 ± 0.0013	0.01508 ± 0.000482	0.1012 ± 0.0084	concordant	96.5 ± 3.1	97.8 ± 8.5
121218-2 no.17	26223	1348	1793783	13010	0.0514 ± 0.0014	0.01501 ± 0.000455	0.1053 ± 0.0070	concordant	96.1 ± 2.9	101.7 ± 7.1
121218-2 no.18	31218	1468	1890633	13712	0.0470 ± 0.0013	0.01696 ± 0.000507	0.1088 ± 0.0070	concordant	108.4 ± 3.3	104.9 ± 7.1
121218-2 no.19	18172	826	1113824	8078	0.0454 ± 0.0012	0.01676 ± 0.000527	0.1039 ± 0.0084	concordant	107.1 ± 3.4	100.4 ± 8.5
121218-2 no.20	10070	1041	646023	4685	0.1033 ± 0.0028	0.01601 ± 0.000547	0.3258 ± 0.0173	-81.5	102.4 ± 3.5	206.7 ± 17.4
121218-2 no.21	29485	2464	1882680	13654	0.0836 ± 0.0022	0.01609 ± 0.000483	0.1834 ± 0.0102	-53.2	102.9 ± 3.1	171.0 ± 10.3
121218-2 no.22(2)	17800	874	1129963	8195	0.0491 ± 0.0013	0.01618 ± 0.000510	0.1084 ± 0.0086	concordant	103.5 ± 3.3	104.5 ± 8.7
121218-2 no.23	12300	648	821909	5961	0.0527 ± 0.0014	0.01537 ± 0.000508	0.1105 ± 0.0099	concordant	98.3 ± 3.3	106.4 ± 10.0
121218-2 no.24	25401	2071	1615421	11716	0.0815 ± 0.0022	0.01615 ± 0.000491	0.1797 ± 0.0105	-48.1	103.3 ± 3.2	167.8 ± 10.7
121218-2 no.25(2)	11249	497	697194	5057	0.0442 ± 0.0012	0.01657 ± 0.000556	0.0999 ± 0.0100	concordant	105.9 ± 3.6	96.6 ± 10.1
121218-2 no.26	15029	678	1006681	7300	0.0451 ± 0.0012	0.01533 ± 0.000493	0.0944 ± 0.0082	concordant	98.1 ± 3.2	91.6 ± 8.3
121218-2 no.27	20339	1057	1365946	9907	0.0520 ± 0.0014	0.01529 ± 0.000475	0.1084 ± 0.0079	concordant	97.8 ± 3.1	104.5 ± 8.0
121218-2 no.28	12780	584	770447	5588	0.0457 ± 0.0012	0.01704 ± 0.000560	0.1062 ± 0.0099	concordant	108.9 ± 3.6	102.5 ± 10.0
121218-2 no.29(2)	17872	877	1259048	9134	0.0491 ± 0.0013	0.01457 ± 0.000459	0.0976 ± 0.0099	concordant	93.3 ± 3.0	94.5 ± 7.8
121218-2 no.30	21128	969	1427509	10353	0.0458 ± 0.0012	0.01520 ± 0.000470	0.0951 ± 0.0072	concordant	97.3 ± 3.0	92.3 ± 7.3
Average ± 2SD												103.3 ± 14.6		132.4 ± 89.0	
Weighted average												102.4 ± 0.6		114.7 ± 1.8	

Disc.* is degree of discordance. Positive and negative values mean left- and right-side significant offsets from concordia line, respectively.

Table 5c-1. U-Pb ages for adopted zircon grains obtained from cluster 1 of Sp. 121218-2.

sample name	Total count				Isotopic ratios						Disc.*	Age (Ma)			
	206Pb	207Pb	238U	235U	207Pb/206Pb	Error 2σ	206Pb/238U	Error 2σ	207Pb/235U	Error 2σ		206Pb/238U	Error 2σ	207Pb/235U	Error 2σ
121218-2 no.8(2)	22805	1161	1539560	11166	0.05091 ± 0.00186		0.01504 ± 0.00046		0.10480 ± 0.00858		concordant	96.3 ± 2.9		101.2 ± 8.7	
121218-2 no.9	15204	834	1027282	7451	0.05485 ± 0.00200		0.01503 ± 0.00048		0.11281 ± 0.01024		concordant	96.2 ± 3.1		108.5 ± 10.3	
121218-2 no.10(2)	8012	425	533831	3872	0.05304 ± 0.00194		0.01524 ± 0.00054		0.11062 ± 0.01278		concordant	97.5 ± 3.5		106.5 ± 12.9	
121218-2 no.16	15658	770	1066452	7735	0.04916 ± 0.00132		0.01508 ± 0.00048		0.10116 ± 0.00839		concordant	96.5 ± 3.1		97.8 ± 8.5	
121218-2 no.17	26223	1348	1793783	13010	0.05139 ± 0.00138		0.01501 ± 0.00046		0.10531 ± 0.00702		concordant	96.1 ± 2.9		101.7 ± 7.1	
121218-2 no.23	12300	648	821909	5961	0.05266 ± 0.00141		0.01537 ± 0.00051		0.11046 ± 0.00989		concordant	98.3 ± 3.3		106.4 ± 10.0	
121218-2 no.26	15029	678	1006581	7300	0.04509 ± 0.00121		0.01533 ± 0.00049		0.09437 ± 0.00824		concordant	98.1 ± 3.2		91.6 ± 8.3	
121218-2 no.27	20339	1057	1365946	9907	0.05195 ± 0.00139		0.01529 ± 0.00047		0.10843 ± 0.00794		concordant	97.8 ± 3.1		104.5 ± 8.0	
121218-2 no.29(2)	17872	877	1259348	9134	0.04905 ± 0.00132		0.01457 ± 0.00046		0.09758 ± 0.00767		concordant	93.3 ± 3.0		94.5 ± 7.8	
121218-2 no.30	21128	969	1427509	10353	0.04585 ± 0.00123		0.01520 ± 0.00047		0.09512 ± 0.00717		concordant	97.3 ± 3.0		92.3 ± 7.3	
Average ± 2SD												96.7 ± 2.9		100.5 ± 12.4	
Weighted average												96.6 ± 1.0		99.4 ± 2.7	

Disc.* is degree of discordance. Positive and negative values mean left- and right-side significant offsets from concordia line, respectively.

Table 5c-2. U-Pb ages for adopted zircon grains obtained from cluster 2 of Sp. 121218-2.

sample name	Total count				Isotopic ratios						Disc.*	Age (Ma)			
	206Pb	207Pb	238U	235U	207Pb/206Pb	Error 2σ	206Pb/238U	Error 2σ	207Pb/235U	Error 2σ		206Pb/238U	Error 2σ	207Pb/235U	Error 2σ
121218-2 no.7(2)	9137	388	565201	4099	0.04246 ± 0.00155		0.01642 ± 0.00057		0.09538 ± 0.01136		concordant	105.0 ± 3.6		92.5 ± 11.5	
121218-2 no.18	31218	1468	1890633	13712	0.04701 ± 0.00126		0.01696 ± 0.00061		0.10882 ± 0.00704		concordant	108.4 ± 3.3		104.9 ± 7.1	
121218-2 no.19	18172	826	1113824	8078	0.04544 ± 0.00122		0.01676 ± 0.00063		0.10391 ± 0.00838		concordant	107.1 ± 3.4		100.4 ± 8.5	
121218-2 no.22(2)	17800	874	1129963	8195	0.04908 ± 0.00132		0.01618 ± 0.00051		0.10838 ± 0.00856		concordant	103.5 ± 3.3		104.5 ± 8.7	
121218-2 no.25(2)	11249	497	697194	5057	0.04415 ± 0.00118		0.01657 ± 0.00056		0.09985 ± 0.00999		concordant	105.9 ± 3.6		96.6 ± 10.1	
121218-2 no.28	12760	584	770447	5588	0.04567 ± 0.00122		0.01704 ± 0.00056		0.10619 ± 0.00993		concordant	108.9 ± 3.6		102.5 ± 10.0	
Average ± 2SD												106.5 ± 4.2		100.2 ± 9.7	
Weighted average												106.5 ± 1.4		101.3 ± 3.7	

Disc.* is degree of discordance. Positive and negative values mean left- and right-side significant offsets from concordia line, respectively.

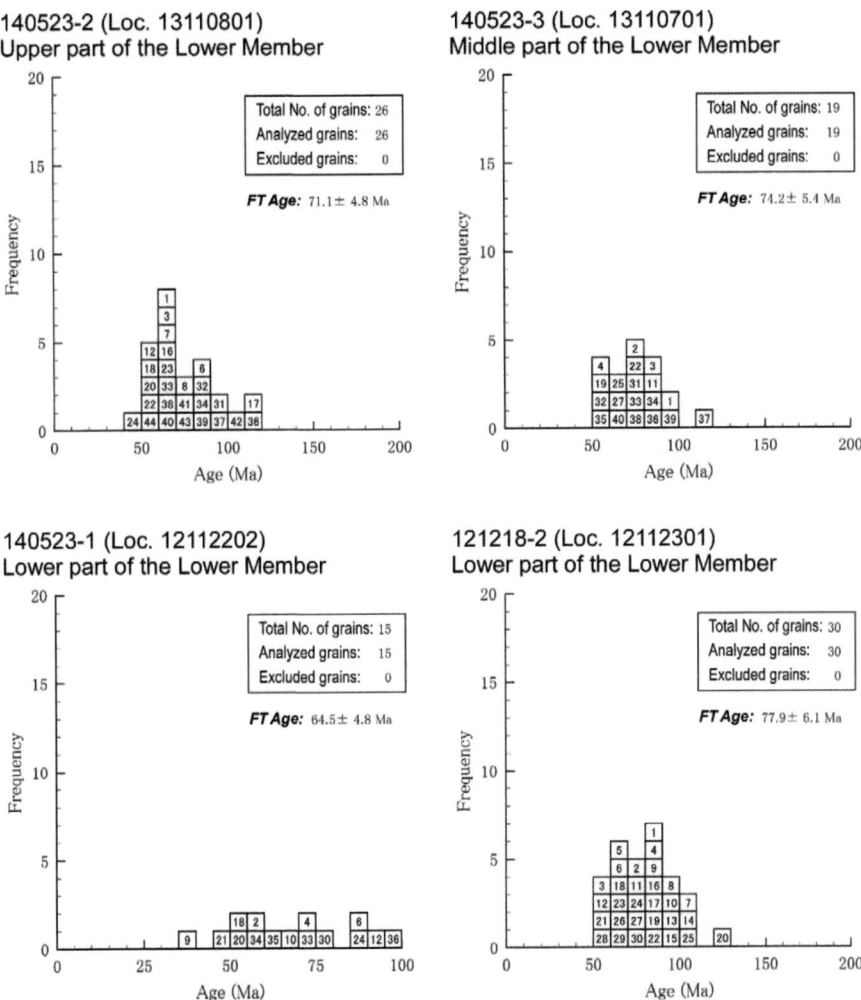

Figure 14a. Results of fission-track (FT) dating of zircons obtained from granite pebbles contained in the Lower Member of the Gunchu Formation. See Figure 6 for locations (Locs. 13110801, 13110701, 12112202, 12112301).

The FT dating results for 30 zircon grains from Sp. 121218-2 are summarized in Table 5a. As shown in Figure 14a, its grain ages are well grouped, averaging 77.9±6.1 Ma (error: 1σ). Figure 14b indicates that the U-Pb ages for this sample are significantly older than its FT ages, and are notably more scattered than the other U-Pb data.

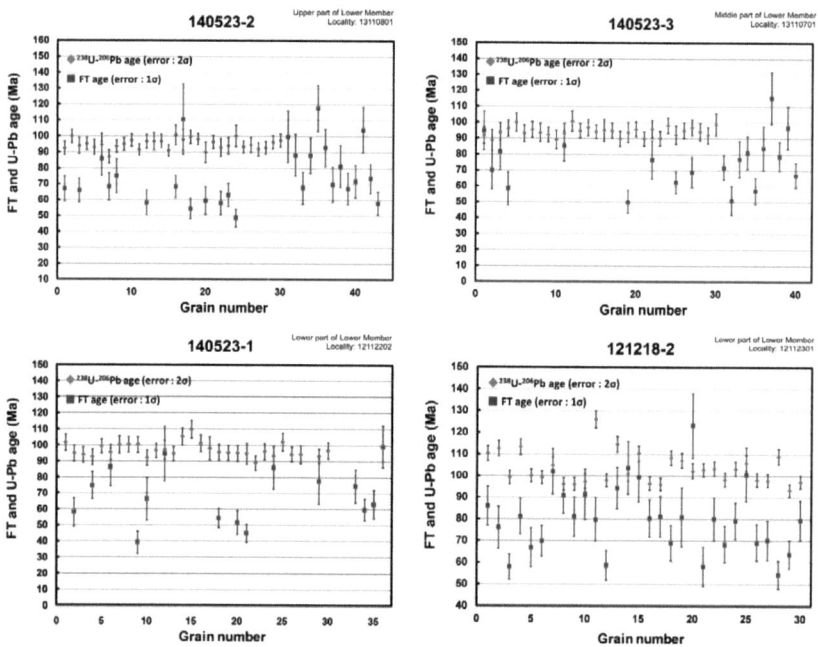

Figure 14b. FT and U-Pb age plots for zircons obtained from granite pebbles contained in the Lower Member of the Gunchu Formation. See Figure 6 for locations (Locs. 13110801, 13110701, 12112202, 12112301).

The U-Pb dating results for total 30 zircon grains from Sp. 121218-2 are summarized in Table 5b. After ruling out zircon grains with discordant ages, 16 concordant grains are obviously divided into two age clusters. Figure 14c shows concordia plots for the two clusters, for which final results are separately presented in Tables 5c-1 and 5c-2, and in two histograms in Figure 14d. The average ages of clusters 1 and 2 are 96.6±1.0 Ma (error: 2σ) and 106.5±1.4 Ma (error: 2σ), respectively.

The FT ages calculated separately for clusters 1 and 2 have indistinguishable values of 77.9±6.3 Ma (error: 1σ) and 78.9±6.7 Ma (error: 1σ), respectively. The closing temperature for U-Pb ages is ca. 900°C in a laboratory or > 700°C in natural conditions, whereas that for zircon FT ages is 240~250°C for heating over a period of 1 million years. This suggests that granite pebbles at the sampled site were derived from two regional plutons, which had been asynchronously emplaced and then exhumed in the same uplift event around the Late Cretaceous.

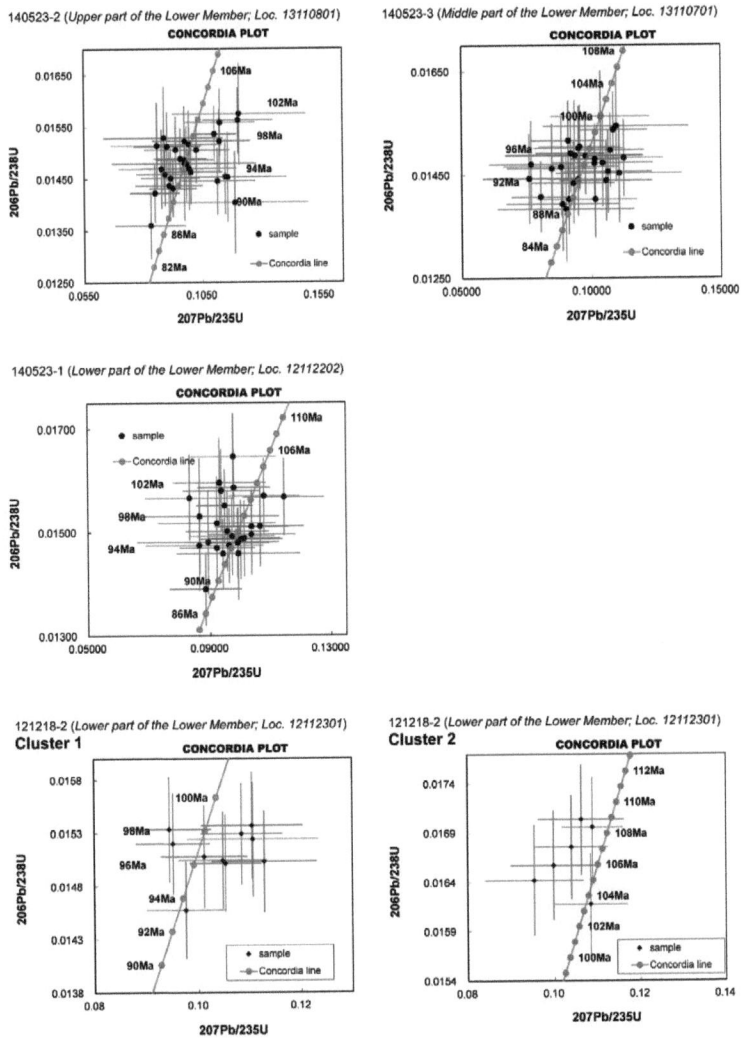

Figure 14c. Concordia plots for U-Pb ages of zircons obtained from granite pebbles contained in the Lower Member of the Gunchu Formation. See Figure 6 for locations (Locs. 13110801, 13110701, 12112202, 12112301).

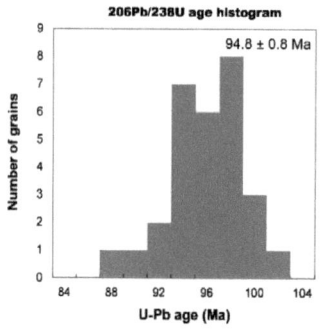

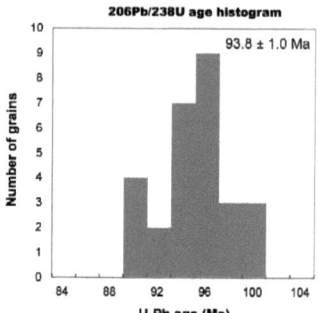

140523-1 (*Lower part of the Lower Member; Loc. 12112202*)

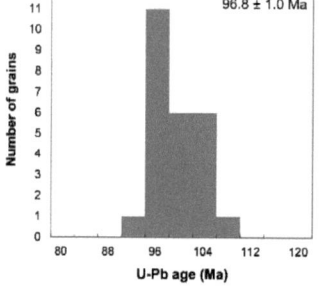

121218-2 (*Lower part of the Lower Member; Loc. 12112301*)
Cluster 1

121218-2 (*Lower part of the Lower Member; Loc. 12112301*)
Cluster 2

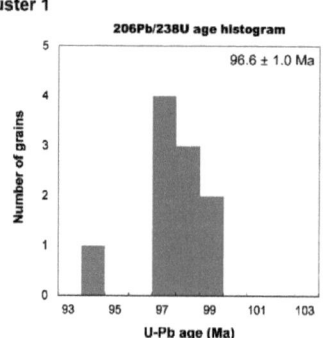

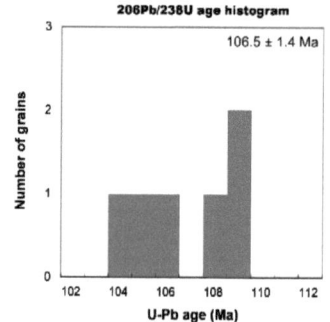

Figure 14d. Results of U-Pb dating of zircons obtained from granite pebbles contained in the Lower Member of the Gunchu Formation. See Figure 6 for locations (Locs. 13110801, 13110701, 12112202, 12112301).

Large Cretaceous granitic plutons have never been found around the study area. Such rocks distribute widely in the southwestern part of Honshu and in northern tip of Shikoku, constituting a metamorphosed terrane of the Inner Zone. Suzuki and Adachi (1998) performed a great deal of CHIME monazite dating for the main exposure of the granites and associated gneisses in Honshu, and obtained ages ranging in from 101 to 85 Ma.

6. Rock magnetic analysis

6.1. Samples

A granulometric study of ferromagnetic minerals was performed to understand the physical properties of the fine sedimentary particles constituting the Gunchu Formation. Samples were taken from 10 sites (GUV01~10) as shown in Figure 6 and Figures 7a to 7d. GUV01, 02, 04 and 05 are sandy silt or fine sand in alternating beds. GUV03, 06 and 07 are fine matrices of gravel beds. GUV08 and 10 are from thick units of sandy silt. GUV09 is the ash layer treated in the tephra analysis (Sp. 13110901).

6.2. Methods

Hysteresis parameters were determined with a Vibrating Sample Magnetometer (MicroMag3900; Princeton Measurements Corporation) at the Center for Advanced Marine Core Research, Kochi University. Each sample, consisting of up to 1 cm^3 of powder, was contained in a gelatin capsule. Figure 15 illustrates the hysteresis loops of the Gunchu Formation. After correction to a linear paramagnetism gradient, we are able to recognize a weak ferromagnetic behavior signature. The 'wasp-waisted' shape of the loops is not observed in the analyzed samples, a fact that indicates the absence of highly coercive components related to a remagnetization event (e.g., Channell and McCabe, 1994).

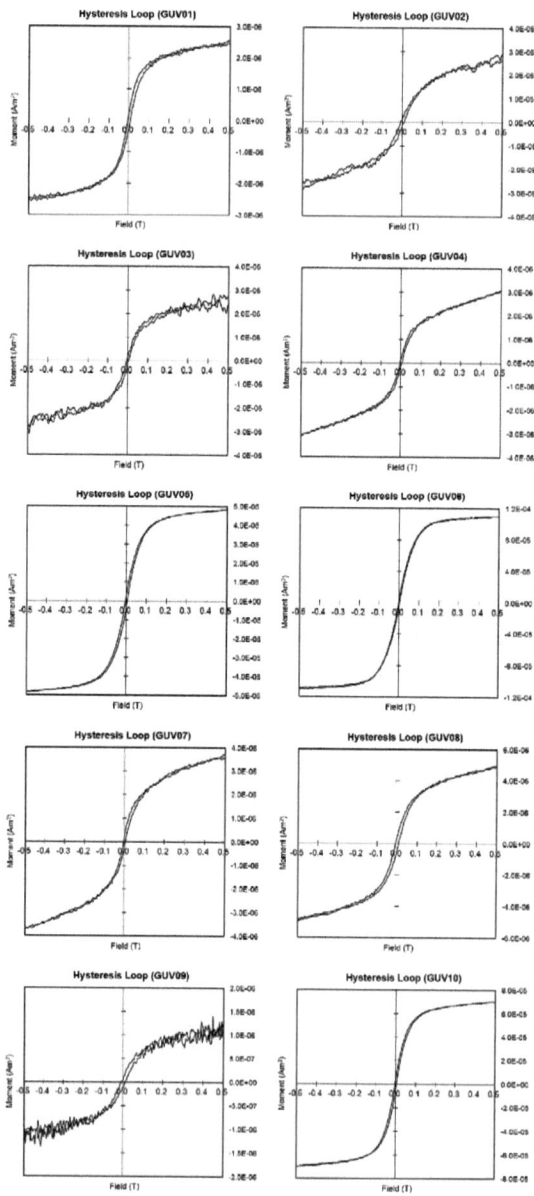

Figure 15. Hysteresis loops for 10 horizons of the Gunchu Formation after correction to a linear paramagnetism gradient. See Figure 6 for locations (GUV01~GUV10).

6.3. Results

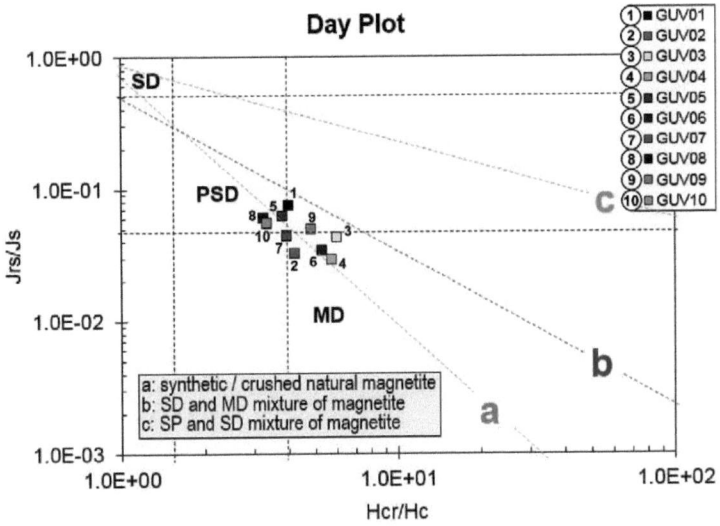

Figure 16. Logarithmic plot of hysteresis parameters (Day et al., 1977) for 10 horizons of the Gunchu Formation. Abbreviations: SD, single domain; PSD, pseudo-single domain; MD, multi-domain. Trends a, b and c originate from synthetic and crushed natural magnetite (Dunlop, 1986), single-domain (SD) and multidomain (MD) mixtures of magnetite (Channell and McCabe, 1994) and superparamagnetic (SP) and SD mixtures of magnetite (Jackson, 1990; Jackson et al., 1993), respectively.

The values for saturation magnetization (Js), saturation remanence (Jrs) and coercive force (Hc) were determined for all samples from their corresponding hysteresis loops. A relatively low Hc (<10 mT) implies the dominance of magnetite as detrital ferromagnetic mineral. Figure 16 shows correlation plots of Jrs/Js vs. Hcr/Hc (Day et al., 1977) using values of the coercivity of remanence (Hcr) obtained through backfield demagnetization experiments. Figure 16 shows that all the data are found around mixture trends of various sizes of magnetite, implying that the effect of secondary mineral precipitation can be seen as minimal. In summary, a series of rock magnetic experiments indicated that the fine sediments of the Gunchu Formation generally contain single-phased low-coercivity ferromagnetic mineral (probably magnetite).

7. Discussion

With all the fruits of the two-year geologic expedition to investigate the conspicuous sedimentary unit now having been presented, we begin to assess their potential as indicators of the complicated tectonic processes along an oblique convergent margin.

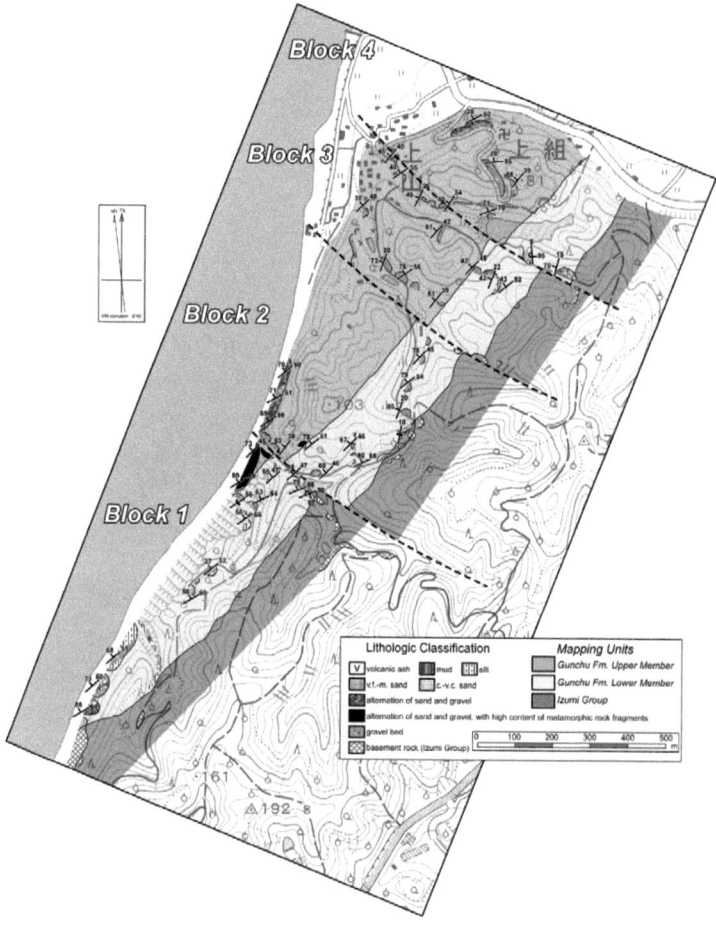

Figure 17. Geologic map of the Gunchu Formation. The base map is a part of the "Kaminada" 1:25,000 topographic map published by the Geographical Survey Institute of Japan.

7.1. Geologic map of the Gunchu Formation

The essence of the study is presented as a geologic map (Figure 17), in which the fault-segmented architecture of the Gunchu Formation is clearly delineated. Bedding attitudes annotated on the map suggest a homoclinal structural buildup during the latter phase of basin formation, as dip angles in the Upper Member are lower than those in the Lower Member. The angular (onlapping) relationship between the sediments and eroded surface of the basement is confirmed for all the segments, which implies syn-depositional tilting of the faulted blocks.

7.2. Migration of depocenter during the Quaternary

As described in previous sections, the Lower Member of the Gunchu Formation is characterized by an influx of a considerable amount of granitic clastics. The nearest exposure of such lithologic units is about 20 km north of the study area (Figure 3). Radiometric dating shows an affinity to the Cretaceous granites in the Inner Zone and the scatter between the ages revealed that gravels in the Lower Member were derived from asynchronously emplaced regional plutons, not from a local intrusive body. These lines of evidence imply that the depocenter of the Iyonada Sea (Figure 3) existed around the area of the present distribution of the Gunchu Formation during its earlier phase of deposition. This situation inevitably led to the formation of a highly asymmetric basin.

In contrast, granite pebbles are absent from the Upper Member, which was first dominantly composed of Sanbagawa metamorphic rocks followed by piled up sandstone gravels of the Izumi Group. This indicates that the Pleistocene depocenter migrated northward in response to an uplift of the Outer Zone, and granites supplied from the Inner Zone were entrapped in the present Iyonada Sea as shown schematically in Figure 18.

7.3. Exhumation of hinterlands

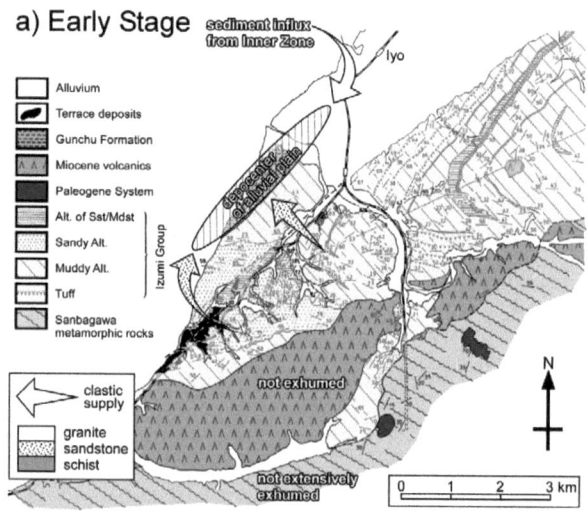

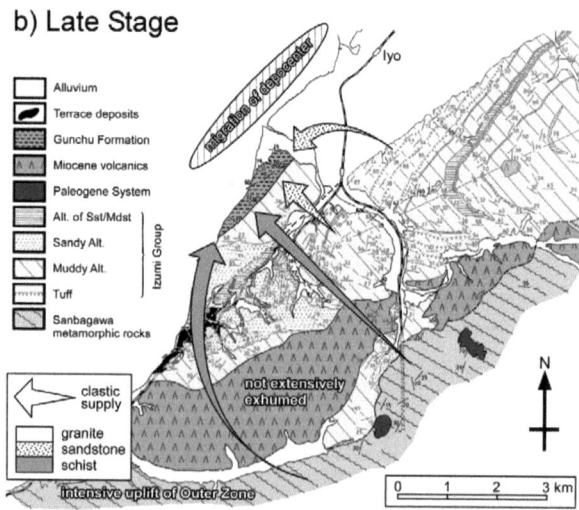

Figure 18. Paleoenvironmental changes during deposition of the Gunchu Formation. Base map is after Takahashi et al. (1990).

A provenance study revealed an intriguing pattern of exhumation of the hinterlands as expressed in Figure 18. Although volcanic clasts remain a minor constituent in the Gunchu Formation, a series of Miocene volcanics is widely distributed between the Gunchu basin and the Sanbagawa terrane as depicted in the figure. Together with the upward progressive dominance of sandstones derived from the Izumi Group, this suggests that the uplift event of the Outer Zone commenced in the early Pleistocene and expanded to the northern areas. The regional tectonic episode eventually caused the nearly vertical tilt of the Gunchu Formation followed by the emergence of the Miocene volcanics that had once been buried during thermal subsidence of southwest Japan after the backarc opening.

The sequence of uplift events seems to reflect a change in the activity mode on the MTLAFS. Combining the development process of the specific structure of the Gunchu Formation described in the next section, the author presents a probable spatiotemporal scenario of deformation on the MTLAFS around the southern margin of the Iyonada Sea.

7.4. Development of unique structure

The geologic map in Figure 17 clearly indicates right-stepping on the faults across the Gunchu basin. This appears to be a discordant pattern, based on a simple model of fault block motions. In a nested fault system developed by dextral activity on the primary transcurrent faults, the faulted blocks should rotate clockwise with sinistral slips on secondary faults as shown in the diagram in Figure 19, which is opposite to the sense of displacement confirmed in this study.

Another important discrepancy becomes apparent with a closer look at the geologic map. It seems that the fault offsets on the boundary between the Lower and Upper Members of the Gunchu Formation tend to be smaller than those of the unconformable interface between the Gunchu Formation and the Izumi Group. This is most obvious on the fault bounding Blocks 1 and 2 where the stratigraphic boundaries were confirmed by an elaborate field survey. Geologic mapping suggests that the same tendency is ubiquitous in the Pleistocene tectonic basin. Provided that we adopt a common nested-fault model, uniform slip on subordinate faults is expected as illustrated in Figure 19. Thus we must devise a new concept to understand the development process of the enigmatic structure.

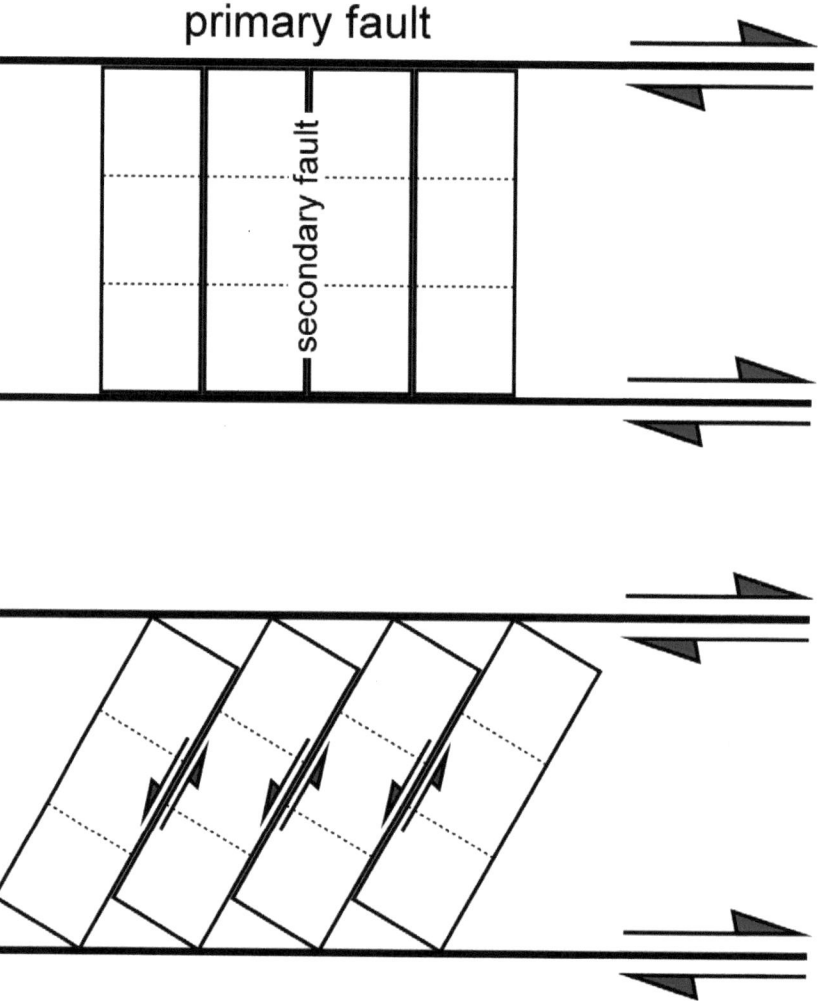

Figure 19. Plan view of general nested-fault system. Primary and secondary faults have opposite sense of lateral motion, and slip on secondary faults is uniform as shown by offsets of dotted lines.

7.5. Geologic reconstruction

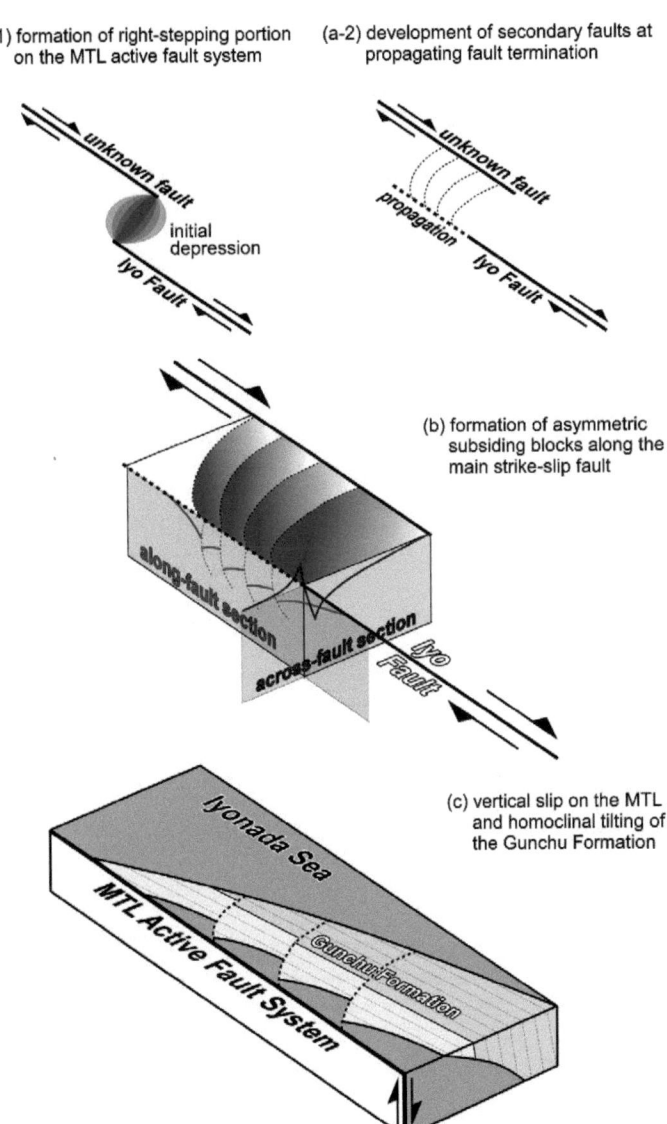

(a-1) formation of right-stepping portion on the MTL active fault system

(a-2) development of secondary faults at propagating fault termination

(b) formation of asymmetric subsiding blocks along the main strike-slip fault

(c) vertical slip on the MTL and homoclinal tilting of the Gunchu Formation

Figure 20. Schematic diagrams showing the probable mechanism of the evolution of the asymmetric Gunchu basin adjoining regional transcurrent faults.

A model of fault-related formation of an asymmetric basin is presented in Figure 20. We assume a strike-slip fault having a right offset from the Iyo Fault. This conceptual structure is in accordance with the broken strands of the MTLAFS around the Iyonada Sea, as reported by Ogawa et al. (1992) and Ikeda et al. (2009). Generally, stepping geometry in a strike-slip fault system induces development of secondary faults connecting the primary fault traces as suggested by Du and Aydin (1995) and Kusumoto et al. (2001). Hence, propagation of the terminal end of the Iyo Fault provoked secondary crustal breaks diverging from the main fault following the formation of an embryotic depression (Figure 20a-1, 2). Fault-bounded sags were systematically tilted, and were deepest at the junctions of the main fault (Iyo) and secondary faults (Figure 20b). Finally vertical movements on the MTLAFS enhanced the steep homoclinal tilting of the Gunchu Formation that had rapidly buried the tectonic asymmetric basin (Figure 20c) and the depocenter migrated seaward. As a result, we currently see an overturned along-fault section of the sedimentary unit.

7.6. Regional tectonic model

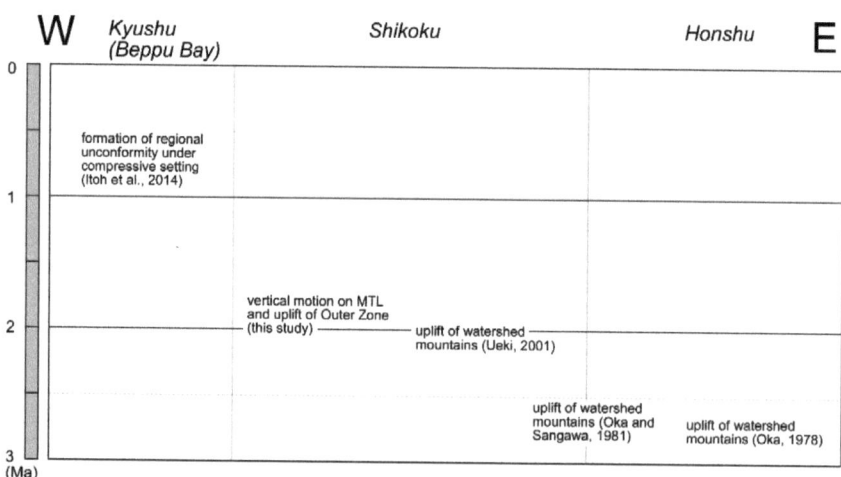

Figure 21. Onset timing of recent compressive events along the Median Tectonic Line active fault system.

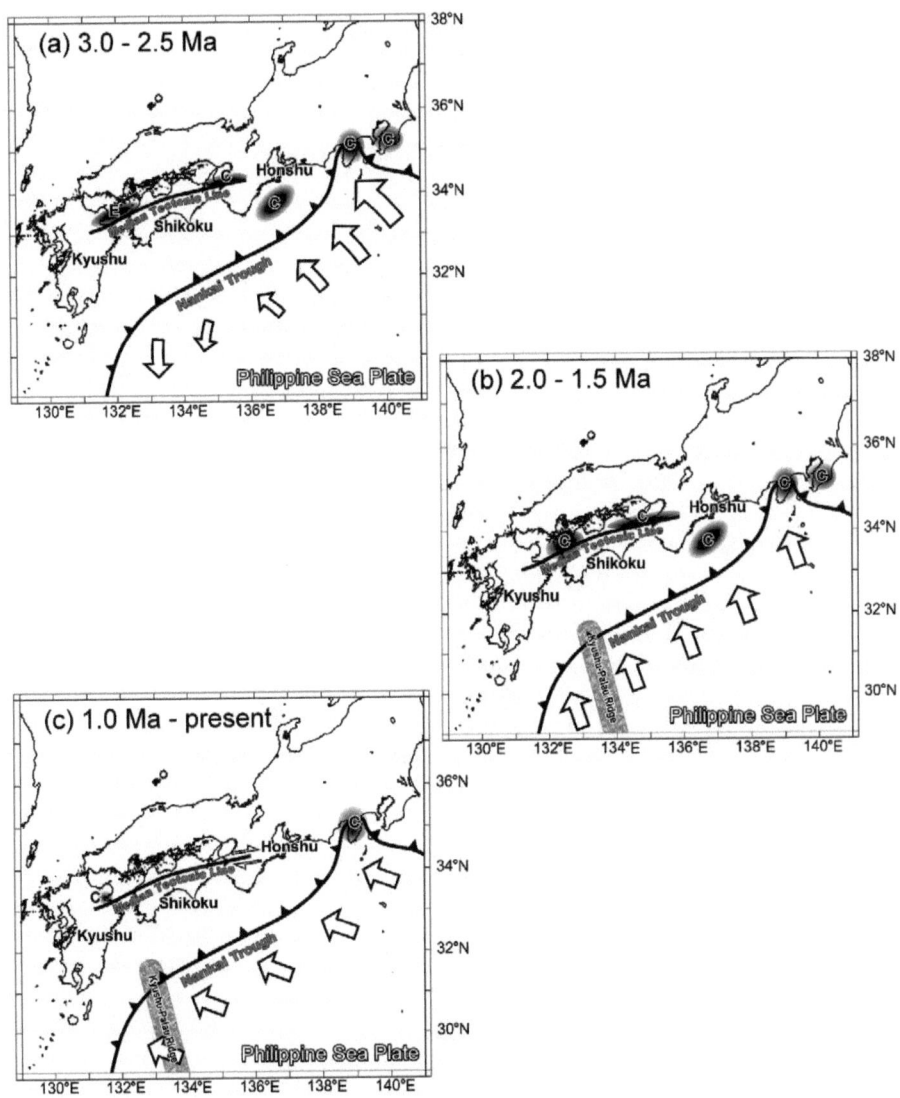

Figure 22. Possible temporal changes of the convergence mode of the Philippine Sea Plate inferred from the deformation modes of southwest Japan. Plate motion is schematically shown by open arrows on the Philippine Sea Plate. Areas of extension and contraction are marked by shaded ovals with letters "E" and "C", respectively. Paleoposition of the Kyushu-Palau Ridge is after Yamazaki and Okamura (1989).

Geologic study of the Gunchu Formation has elucidated a contractional event intervening in active dextral movements on the MTLAFS around the northwestern part of Shikoku. Similar tectonic episodes have been reported in some areas in southwest Japan. From east to west, Honshu, Awaji (an island between Honshu and Shikoku), Shikoku, and central Kyushu suffered contraction in the Pleistocene. Their spatiotemporal relationship is examined as follows.

Oka (1978) studied the emergence of the watershed Izumi Mountains along the MTLAFS, passing through the Kii Peninsula of Honshu, by investigating the gravel composition of the Pleistocene strata in the Inner Zone. He pointed out that influx of schist gravels from metamorphosed terranes in the Outer Zone ceased around the horizon of the Mizuma II volcanic ash, which was assigned to 3.0~2.5 Ma by Satoguchi and Nagahashi (2012). Oka and Sangawa (1981) described metamorphic pebbles in the Pleistocene sediments distributed within Awaji Island on the Inner Zone. They attributed a limited stratigraphic occurrence of schists to an uplift of the watershed Yuzuruha Mountains along the southern coast, which should have been caused by vertical activity on the MTLAFS running off the island. According to Mizuno (1992), the sediments can be correlated with the lower part of the Pleistocene Osaka Group in Honshu. As for the eastern part of Shikoku, Ueki (2001) also argued for uplift of the dividing range along the MTLAFS based on changes in the provenance of clastics of the Mitoyo Group resting in the Inner Zone. Mizuno (1992) assigned the sedimentary unit to the middle part of the Osaka Group. According to his tephrostratigraphy, the Mitoyo Group is correlatable with the Gunchu Formation. In central Kyushu, Itoh et al. (2014) executed a tectono-stratigraphic study around the Beppu Bay utilizing reflection seismic data. Based on geologic evidence in the surrounding areas, they assigned a regional contractional event to the unconformity between the Oita and Sekinan Groups (0.7 Ma).

The spatiotemporal positions of the Pleistocene tectonic events in southwest Japan are summarized in Figure 21. It seems that the rise of the contraction is younger going westward. Itoh et al. (2014) argued for an older phase of the growth of the contractional domain, and attributed the transient tectonic regimes to changing convergence modes of the Philippine Sea Plate. As shown in insets a and c of Figure 22, they assumed episodic shifts of the Euler pole of the Philippine Sea/Eurasian Plates. The present study has reconstructed a more precise deformation history of the convergent margin, and the author advocates an in-between phase of northerly subduction of the Philippine Sea Plate and widespread contraction (Figure 22b), which corresponds to the period of north-northwestward subduction originally proposed by Nakamura et al. (1987).

There may be another explanation for the intensive structural buildup around the Gunchu basin. Figure 22 shows the approximate position of the Kyushu-Palau Ridge. The tip of the remnant arc on the Philippine Sea Plate reaches the axis of the Nankai Trough, and its geomagnetic anomaly (Figure 1) implies that an unrevealed northern extension is underthrusting into the accretionary prism of the forearc of southwest Japan. According to a structural restoration of the forearc wedge (Yamazaki and Okamura, 1989), western Shikoku was located in front of the northerly indenting obstacle on the oceanic plate in the early Pleistocene. Such a condition resulted in strong compressive stress and vertical motions on the MTLAFS around the study area. This may also solve the mechanism of massive uplift and erosion of Shikoku Island mentioned before, which supplied enough clastics to bury the gigantic tectonic depression in the Iyonada Sea.

8. Conclusions

A well-organized geologic survey of a tectonic basin adjacent to an arc-bisecting fault along a convergent margin was conducted on the Philippine Sea/Eurasian Plate boundary. Multidisciplinary analyses of a Pleistocene alluvial unit, the Gunchu Formation, on the Median Tectonic Line active fault system (MTLAFS) have revealed the following points.

1) The Gunchu Formation is a cycle of gravel-dominant nonmarine deposits showing an upward-coarsening sequence. It rapidly buried an elongate asymmetric basin along the MTLAFS around 2 Ma.

2) The Gunchu Formation is divided into the Lower and Upper Members. The former contains considerable amounts of granite pebbles derived from the Inner Zone to the

north (backarc side) of the MTLAFS, while the latter is largely composed of clastics from the southern (forearc side) terranes, such as low T/P metamorphic rocks to the south of the MTLAFS (Outer Zone) and a Cretaceous sandstone.

3) The Gunchu Formation exhibits a basin-ward steep homoclinal tilting, which developed during the later phase of its formation. Structural buildup reflects progressive uplift and exhumation of the Outer Zone, which caused the drastic change in the provenance of clastic material and migration of depocenter of the elongate basin.

4) Simultaneously with the formation of the homoclinal structure, the tectonic basin was divided by some secondary faults. The fault-bounded blocks systematically tilted, having their deepest portion at junctions where the main fault of the MTLAFS met diverging secondary faults.

5) The conspicuous structural style is understood as the result of the process of secondary faults developing as a result of the propagation of the terminal end of the dextral MTLAFS, which was followed by vertical movements on the main fault system.

6) Quaternary contractional tectonic episodes accompanied by vertical activity on the MTLAFS were ubiquitous in southwest Japan, and their commencement tends to be younger going westward along the convergent margin. This tendency is related to a jump of the Euler pole of the Philippine Sea/Eurasian Plates. Normal subduction of the oceanic plate resulted in prevailing compressive stress in the middle Pleistocene. Strong deformation of the Gunchu Formation may have been enhanced by underthrusting of an obstacle on the Philippine Sea Plate, namely, a remnant arc of the Kyushu-Palau Ridge.

Acknowledgements

The author is grateful to K. Takemura for his thoughtful comments during the course of this study. Thanks are also due to Y. Yamamoto for the use of the rock magnetic laboratory at Kochi University. The field survey was financially supported by the Integrated Research Project for Active Fault Systems of the Ministry of Education, Culture, Sports, Science and Technology (MEXT), Japan.

References

Channell, J.E.T., McCabe, C., 1994. Comparison of magnetic hysteresis parameters of unremagnetized and remagnetized limestones. Journal of Geophysical Research 99, 4613-4623.

Danhara, T., Iwano, H., 2009. Determination of zeta values for fission-track age calibration using thermal neutron irradiation at the JRR-3 reactor of JAEA, Japan. Journal of Geological Society of Japan 115, 141-145.

Davis, P.J., Rabinowitz, P., 2007. Methods of Numerical Integration, 2nd Edition. Dover, New York.

Day, R., Fuller, M., Schmidt, V.A., 1977. Hysteresis properties of titanomagnetites: grain-size and compositional dependence. Physics of the Earth and Planetary Interiors 13, 260-267.

Du, Y., Aydin, A., 1995. Shear fracture patterns and connectivity at geometric complexities along strike-slip faults. Journal of Geophysical Research 100, 18093-18102.

Dunlop, D.J., 1986. Hysteresis properties of magnetite and their dependence on particle size: a test of pseudo-single-domain remanence models. Journal of Geophysical Research 91, 9569-9584.

Engebretson, D.C., Cox, A., Gordon, R.C., 1985. Relative motions between oceanic and continental plates in the Pacific Basin. Geological Society of America Special Paper 206, 1-59.

Fitch, T.J., 1972. Plate convergence, transcurrent faults, and internal deformation adjacent to southeast Asia and the western Pacific. Journal of Geophysical Research 77, 4432-4460.

Geological Survey of Japan, AIST (Ed.), 2012. Seamless Digital Geological Map of Japan 1:200,000 (July 3, 2012 Version), Research Information Database DB084. Geological Survey of Japan, AIST (National Institute of Advanced Industrial Science and Technology), Tsukuba.

Geological Survey of Japan, AIST, CCOP (Coordinating Committee for Coastal and

Offshore Geoscience Programmes in East and Southeast Asia) (Eds.), 2002. Magnetic Anomaly Map of East Asia 1:4,000,000, CD-ROM Version (2nd Edition), Digital Geoscience Map P-3. Geological Survey of Japan, AIST (National Institute of Advanced Industrial Science and Technology), Tsukuba.

Ikeda, M., Toda, S., Kobayashi, S., Ohno, Y., Nishizaka, N., Ohno, I., 2009. Tectonic model and fault segmentation of the Median Tectonic Line active fault system on Shikoku, Japan. Tectonics 28, TC5006, doi:10.1029/2008TC002349.

Itoh, Y. (Ed.), 2013. Mechanism of Sedimentary Basin Formation - Multidisciplinary Approach on Active Plate Margins. InTech, Rijeka (Croatia), http://dx.doi.org/10.5772/50016.

Itoh, Y., Takemura, K., 1993. Quaternary geomorphic trends within Southwest Japan: extensive wrench deformation related to transcurrent motions of the Median Tectonic Line. Tectonophysics 227, 95-104.

Itoh, Y., Kusumoto, S., Takemura, K., 2013. Characteristic basin formation at terminations of a large transcurrent fault - basin configuration based on gravity and geomagnetic data. In: Itoh, Y. (Ed.) Mechanism of Sedimentary Basin Formation - Multidisciplinary Approach on Active Plate Margins. InTech, Rijeka (Croatia), http://dx.doi.org/10.5772/56702.

Itoh, Y., Kusumoto, S., Takemura, K., 2014. Evolutionary process of Beppu Bay in central Kyushu, Japan: a quantitative study of the basin-forming process controlled by plate convergence modes. Earth, Planets and Space 66, 74, http://www.earth-planets-space.com/content/66/1/74.

Itoh, Y., Takemura, K., Kamata, H., 1998. History of basin formation and tectonic evolution at the termination of a large transcurrent fault system: deformation mode of central Kyushu, Japan. Tectonophysics 284, 135-150.

Itoh, Y., Tsutsumi, H., Yamamoto, H., Arato, H., 2002. Active right-lateral strike-slip fault zone along the southern margin of the Japan Sea. Tectonophysics 351, 301-314.

Itoh, Y., Uno, K., Arato, H., 2006. Seismic evidence of divergent rifting and subsequent deformation in the southern Japan Sea, and a Cenozoic tectonic synthesis of the eastern Eurasian margin. Journal of Asian Earth Sciences 27, 933-942.

Jackson, M., 1990. Diagenetic sources of stable remanence in remagnetized Paleozoic cratonic carbonates: a rock magnetic study. Journal of Geophysical Research 95, 2753-2761.

Jackson, M., Rochette, P., Fillion, G., Banerjee, S., Marvin, J., 1993. Rock magnetism of remagnetized Paleozoic carbonates: low-temperature behavior and susceptibility characteristics. Journal of Geophysical Research 98, 6217-6225.

Kitabayashi, E., Danhara, T., Iwano, H., 2012. Fission-track age of zircon from volcanic ash layer of the Gunchu Formation in Iyo City, Ehime Prefecture in Shikoku, Japan. Journal of the Geological Society of Oita 18, 61-64.

Kobayashi, K., Nakada, M., 1978. Magnetic anomalies and tectonic evolution of the Shikoku inter-arc basin. Journal of Physics of the Earth, supplement 26, S391-S402.

Koizumi, K., Fujimoto, H., Inokuchi, H., Uchitsu, M., Kono, Y., 1994. Marine gravity measurements over the Seto Inland Sea, western Japan. Journal of Geodetic Society of Japan 40, 333-345.

Kono, Y., Nishiyama, Y., Inokuchi, H., 2001. Gravity measurements on islands in eastern part of the Seto Inland Sea, western Japan - a high gravity anomaly belt. Journal of Geodetic Society of Japan 47, 649-658.

Kusumoto, S., Fukuda, Y., Takemura, K., Takemoto, S., 2001. Forming mechanism of the sedimentary basin at the termination of the right-lateral left-stepping faults and tectonics around Osaka Bay. Journal of Geography 110, 32-43.

Mizuno, K., 1987. Preliminary report on the Plio-Pleistocene sediments distributed along the Median Tectonic Line in and around Shikoku, Japan. Bulletin of the Geological Survey of Japan 38, 171-190.

Mizuno, K., 1992. Age and tectonic development of the Plio-Pleistocene sedimentary basins along the Median Tectonic Line, Southwest Japan. Memoirs of the Geological Society of Japan 40, 1-14.

Nagai, K., 1957. Geology of the Ehime Prefecture. Tomoeya, Matsuyama.

Nakamura, K., Renard, V., Angelier, J., Azema, J., Bourgois, J., Deplus, C., Fujioka, K., Hamano, Y., Huchon, P., Kinoshita, H., Labaume, P., Ogawa, Y., Seno, T., Takeuchi, A., Tanahashi, M., Uchiyama, A., Vigneresse, J.L., 1987. Oblique and near collision subduction, Sagami and Suruga Troughs - preliminary results of the

French-Japanese 1984 Kaiko cruise, Leg 2. Earth and Planetary Science Letters 83, 229-242.

Nakata, T., Imaizumi, T. (Eds.), 2002. Digital Active Fault Map of Japan (in Japanese). University of Tokyo Press, Tokyo, 60pp. and 2 DVDs.

Nishimura, S., Hashimoto, M., 2006. A model with rigid rotations and slip deficits for the GPS-derived velocity field in Southwest Japan. Tectonophysics 421, 187-207.

Ogawa, M., Okamura, M., Shimazaki, K., Nakata, T., Chida, N., Nakamura, T., Miyatake, T., Maemoku, H., Tsutsumi, H., 1992. Holocene activity on a submarine active fault system of the Median Tectonic Line beneath the northeastern part of Iyonada, the Inland Sea, Southwest Japan. Memoir of the Geological Society of Japan 40, 75-97.

Ohno, I., Kono, Y., Fujimoto, H., Koizumi, K., 1994. Gravity anomaly in and around the western Seto Inland Sea and subsurface structure of negative anomaly belt. Zisin (Bulletin of the Seismological Society of Japan) 47, 395-401.

Oka, Y., 1978. The formation of the Izumi Range and the Osaka Group. Quaternary Research 16, 201-210.

Oka, Y., Sangawa, A., 1981. The formation of the sedimentary basin in the east of Inland Sea and the uplift of the Awaji Island, Japan. Journal of Geography 90, 393-409.

Okada, A., Tsutsumi, H., Nakata, T., Goto, H., Niwa, S., 1998. Active Fault Map in Urban Area 1:25,000 - Gunchu. Geographical Survey Institute, Tsukuba.

Okino, K., Shimakawa, Y., Nagaoka, S., 1994. Evolution of the Shikoku Basin. Journal of Geomagnetism and Geoelectricity 46, 463-479.

Otofuji, Y., Matsuda, T., 1987. Amount of clockwise rotation of Southwest Japan - fan shape opening of the southwestern part of the Japan Sea. Earth and Planetary Science Letters 85, 289-301.

Otofuji, Y., Hayashida, A., Torii, M., 1985. When was the Japan Sea opened?: paleomagnetic evidence from Southwest Japan. In: Nasu, N., Uyeda, S., Kushiro, I., Kobayashi, K., Kagami, H. (Eds.) Formation of Active Ocean Margins. Terra Publishing Co., Tokyo, pp.551-566.

Satoguchi, Y., Nagahashi, Y., 2012. Tephrostratigraphy of the Pliocene to Middle

Pleistocene Series in Honshu and Kyushu Islands, Japan. Island Arc 21, 149-169.

Seno, T., Maruyama, S., 1984. Paleogeographic reconstruction and origin of the Philippine Sea. Tectonophysics 102, 53-84.

Seno, T., Stein, S., Gripp, A.E., 1993. A model for the motion of the Philippine Sea Plate consistent with NUVEL-1 and geological data. Journal of Geophysical Research 98, 17941-17948.

Suzuki, K., Adachi, M., 1998. Denudation history of the high T/P Ryoke metamorphic belt, southwest Japan: constraints from CHIME monazite ages of gneisses and granitoids. Journal of Metamorphic Geology 16, 23-37.

Takahashi, J., Kashima, N., 1985. On the Gunchu Formation at the Mori Coast Iyo City, Ehime Prefecture. Memoirs of Faculty of Education, Ehime University, Natural Science 5, 19-29.

Takahashi, J., Yamasaki, T., Yokota, Y., Kawanishi, J., Inoue, S., 1990. Geology of the Iyo City - Futami Town area, Ehime Prefecture. Memoirs of Faculty of Education, Ehime University, Natural Science 10, 19-29.

Takano, O., Itoh, Y., Kusumoto, S., 2013. Variation in forearc basin configuration and basin-filling depositional systems as a function of trench slope break development and strike-slip movement: examples from the Cenozoic Ishikari - Sanriku-Oki and Tokai-Oki - Kumano-Nada Forearc Basins, Japan. In: Itoh, Y. (Ed.) Mechanism of Sedimentary Basin Formation - Multidisciplinary Approach on Active Plate Margins. InTech, Rijeka (Croatia), http://dx.doi.org/10.5772/56751.

Talwani, M., Lamar, W.J., Landisman, M., 1959. Rapid gravity computations for two-dimensional bodies with application to the Mendocino Submarine Fracture Zone. Journal of Geophysical Research 64, 49-59.

Ueki, T., 2001. The Mitoyo Group along the northern flank of Asan Mountains, central Kagawa Prefecture, southwest Japan: its distribution, stratigraphy, lithofacies and depositional history. Journal of Geography 110, 708-724.

Yamazaki, T., Okamura, Y., 1989. Subducting seamounts and deformation of overriding forearc wedges around Japan. Tectonophysics 160, 207-229.

Yoshikawa, S., 1976. The volcanic ash layers of the Osaka Group. Journal of Geological Society of Japan 82, 497-515.

Yusa, Y., Takemura, K., Kitaoka, K., Kamiyama, K., Horie, S., Nakagawa, I., Kobayashi, Y., Kubotera, A., Sudo, Y., Ikawa, T., Asada, M., 1992. Subsurface structure of Beppu Bay (Kyushu, Japan) by seismic reflection and gravity survey. Zisin (Bulletin of Seismological Society of Japan) 45, 199-212.